SUSTAINABILITY

WITHDRAWN **THE BASICS**

Sustainability: The Basics is a concise and engaging introduction to the central concepts in sustainability studies, which introduces the key debates around the preservation of both society and the environment. It answers such questions as:

- What do we mean by sustainability?
- What are the core principles of sustainability?
- How is sustainability measured and assessed?
- Who decides who and what should be sustained?
- What can we learn from the collapse of previous civilizations?

With contemporary and historic case studies, suggestions for further reading and a glossary of key terms, *Sustainability: The Basics* is an essential read for anyone who wants to know more about key issues in sustainability.

Peter Jacques is Associate Professor in the Department of Political Science at the University of Central Florida and is currently the Sustainability Faculty Fellow.

D0540236

The Basics

SUSTAINABILITY

THE BASICS

Peter Jacques

Routledge
Taylor & Francis Group

LONDON AND NEW YORK

First published 2015
by Routledge
2 Park Square, Milton Park, Abingdon, Oxon OX14 4RN

and by Routledge
711 Third Avenue, New York, NY 10017

Routledge is an imprint of the Taylor & Francis Group, an informa business

British Library Cataloguing in Publication Data
A catalogue record for this book is available from the British Library

Library of Congress Cataloging in Publication Data
Jacques, Peter.
Sustainability : the basics / Peter Jacques.
pages cm
Includes bibliographical references and index.
1. Sustainability--Philosophy. I. Title.
GE195.J34 2014
338.9'2701--dc23
2014003900

ISBN: 978-0-415-60847-3 (hbk)
ISBN: 978-0-415-60848-0 (pbk)
ISBN: 978-1-315-76278-4 (ebk)

Typeset in Bembo
by Taylor & Francis Books

To my beloved, Racine, meus diligo vobis mos permaneo quoad gramen est viridis quod ventus verbera.

CONTENTS

LIST OF ILLUSTRATIONS

FIGURES

TABLES

BOXES

ACKNOWLEDGMENTS

A book typically has many authors, and I must acknowledge several people who were essential to producing this text, none of whom is to blame for any errors. Students in my course, Sustainability, at the University of Central Florida, were exposed to and provided extensive comments on the chapters in this volume, and I owe each of them my gratitude. In addition, founding members of my Political Ecology Lab also read, edited, and provided helpful comments or provided support for those who did: Greg Norris, Chelsea Piner, Cheyenne Canon, Paul-Henry Blanchet, Charlene Kormondy, Sebastian Sarria, and Clayton Besaw. Finally, my beautiful wife, Racine, provided support for me to finish, especially during my time immobilized with a ruptured Achilles tendon (from a sporting game of rugby!) where she had to drive me absolutely everywhere, carry everything for me, and take care of me while I remained powerless to do anything but write and finish this book.

INTRODUCTION

In many ways, this book is an unwise project. To attempt a brief introduction that allows for only limited and basic exploration of the complex, contested, and uncompromising set of ideas bound in the notion of sustainability is truly a fool's errand. Thus, dear reader, this book is only a beginning, and a limited one at that, but a beginning is necessary in its own right. If this book is able to communicate anything, it is that the current trajectory of human consumption of the Earth's ecosystems is both unsustainable because it cannot continue without causing deep social crisis from systemic exhaustion and because it is already creating social crisis through deeply unjust distribution of wealth and wellbeing. Our work must begin now to turn these trends around if we want to continue the human project, and in this way, a primer on the issues, fault lines, and solutions is necessary. Indeed, sometimes students who take my course in Sustainability here at the University of Central Florida often tell me it should be required coursework for every student, as something like a core civic literacy. Naturally, I like to hear that the coursework is intellectually meaningful, but also that teaching and learning have an effect.

This book is about global human sustainability, and the basic debates and problems that sustaining a thriving human species involves. Thus, when I use the embattled word, "we," I mean the human population, though there is no pretense that the world is united in its goals, identities, responsibilities, or impacts. In this text there is no commitment to sustaining any one form of human organization, such as the system of nation-states, or world capitalism, but focus on what threatens and potentially promises a thriving global human society. This ambivalence to the kind of systems we sustain comes from a real sense that we should only sustain systems we find "good" and worthy of sustaining, and new possibilities come from the actual collapse of some systems, especially to the slaves of those systems.

This global society exists through various organized systems that extend imperfectly and unevenly in the twenty-first century. At this point in history, two things are global but unevenly experienced around the world. One, is that the nation-state is the government of last resort, and to have legitimacy in international society, a government must be a nation-state. This specific institution emerged over a long gestation from the absolute monarchs of sixteenth-century Europe. Second, is the economic system of market capitalism, which has evolved from mercantile capitalism of the colonial era, to a liberal capitalism that involved the welfare state social protections, to the current form of neoliberal capitalism that mainly transfers power from the social sphere to the private sphere of firms. In addition, a set of Western values has spread as the dominant social paradigm in world trade arrangements and international agreements, again unevenly but effectively, around the world. These values are those we find in Enlightenment liberalism that focus on individual liberty, the pre-eminence of private property, a faith in industrial science and technology, a faith in future abundance, a preference for limited government and deregulation (Dunlap and Van Liere, 1984). These are all values consistent with the current system of neoliberalism, which divests power and social control from public arenas such as government and civil society to economic arenas and actors such as corporations (Centeno and Cohen, 2012). Consequently, references to the current sociopolitical arrangements or the current world-system in this book refer to these uneven systems of world capitalism, nation-states that have specific roles in capitalism, and the social values that tend to favor global capitalism.

"Thriving" here implies that global human sustainability is not met if there are whole groups of people whom are not "sustained."

Perhaps such a project is impossible, but it is certainly not possible if we never set such conditions as goals. If we imagine the network of societies around the world constitute a world civilization, then a thriving world civilization has failed to this point. Even as some improvements have been made for some populations, including longer lifespans, almost half of the world continues to live with some form of malnutrition. Indeed, radical poverty is organized, not random. Around the world, poverty and hunger exist with alarming consistency in former colonies of Europe in Asia, Latin America, and Sub-Saharan Africa (Friedman and McMichael, 1989; Ponting, 2007; Vernon, 2007). Thus the patterns of hunger have a strong correlation to historical colonialism, where most countries struggling with systemic hunger are in Asia, Latin America (less so now), and Sub-Saharan Africa—all of which had vast areas under colonial rule. In part, this is because this colonial history removed so much capital—an asset that provides income of some kind—from these areas and put this wealth in the core imperial powers. Imperialism removed local institutional capacities by fundamentally changing the social rules of the colonies, changing the infrastructure used in these societies, and removing people and natural resources, all of which altered the course of history for the colonized areas (Bagchi, 2005). While some scholars believe that world poverty is due to underdeveloped countries' failure to follow Western patterns of triumph and (ironically) exceptionalism (Ferguson, 2011), others believe that this critical social problem for sustainability is found not in the failure of poorer societies to perform well, but in their suppression. However, these arrangements are fundamentally changing with the rise of China, India, Russia, and Brazil and other regional powers that are accumulating capital through growing economies and establishing growing middle-class consumer societies. If only a few consumer societies have driven ecosystems to the brink, more consumption will drive more ecosystem problems; however, poor countries fairly demand that it is "their turn" and Western powers have little moral justification for imposing limits after they have eaten at the table, so to speak.

Indeed, some of the most pressing United Nations' goals for development, the Millennium Development Goals (MDG), have had some success. One goal was to halve the amount of people living on less than $1.25 a day by 2015, but it was met in 2010; as was the goal to halve the proportion of people living without "improved" water sources. On the other hand, around 870 million people are severely undernourished as of this writing, gender inequality persists with life and death consequences (these involve the ability of women to have access to equal medical care, education, food, and political rights), and more than a third of the world's population lacks basic sanitation—which is a merciless source of disease and contributes to difficulties in meeting the MDG for reducing the amount of children who die before the age of five. Indeed, the number of children who die in the first month of life is increasing and childhood mortality is twice as likely to affect poor, rather than more affluent, households.

In addition, meeting the challenges of Goal 7 of the MDG of effectively integrating sustainable development into policies around the world and reversing ecological decline is getting much farther away. Experts agree that, "Biodiversity is in crisis. The combined effects of habitat loss, exploitation, invasive species and climate change imperil species in all regions on Earth"(Morris, 2011), while forests are disappearing, and marine ecosystems are experiencing multiple crises from acidification to biological homogenization. This is not surprising since experts tell us that we really need to protect at least half of the world's land for conservation purposes if we are to really address biodiversity loss, and keep the ecological goods and services that make human life possible to begin with intact (Noss et al., 2012).

Through energy production, transportation, and other, mostly industrialized, economic activities, the world has released increased carbon dioxide levels that force climate change and global warming by almost 50 percent since 1990, and 40 percent since the **Industrial Revolution**, causing a cascade of ecological changes observable now from warming ocean water, loss of Arctic sea ice, drying soils and hotter temperatures that undermine crop harvests, to increasing extreme events such as drought, hurricanes, and forest fires. It is also clear that each of these elements interacts with other ecosystem and social changes in malicious ways. Warming conditions are likely to hurt the world's poorest people the hardest, such as in Sub-Saharan Africa, by reducing the crops and increasing disease such as malaria.

Thus, if the goal is a thriving global human population, there is a lot of ground to cover at a time when challenges, especially ecological degradation, continue to grow. For this reason, United Nations' assessments have noted that the other MDGs are not attainable without achieving Goal 7 and building the integrity of Earth's ecological life support systems (Millennium Ecosystem Assessment, 2005c). Yet, human sustainability is not automatically determined by the problems we face but by the resilience we build with each other.

Consequently, an inquiry into human sustainability requires tough ethical questions, especially how other people and non-humans are treated. Sustainability also means we can't have it all. Painful trade-offs with economic growth, social equality, and environmental protections are all part of the discussion, and global humanity itself is divided on how to privilege any one of these interests. Affluent groups want continued economic growth, while the poor may want more equity, opportunity, and justice. An interest in environmental protection spans varied groups, some of whom require more protected water than a pretty view, while still others sympathize for animals or fear a future without polar bears, tigers, or fish. As I finished writing this book, the western black rhino was declared extinct by the International Union for Conservation of Nature and Natural Resources (IUCN), and if this is true, its mammoth tread will never again press upon the Earth and we are forced to explain what a western black rhino was to our grandchildren through history books. The processes that rendered the rhino and other organisms lost to the web of life on Earth are nowhere deemed sustainable, which means that humanity must end this death march if we want a livable planet.

The notion of **sustainability** is at once quite simple and stupendously complex. To sustain something is to keep it going, which is conceptually simple. If I have an orange tree and I want to keep harvesting oranges, I need to foster the tree's health, protect it from being cut down by others, and think ahead by planting successors. But if we are talking about the global sustainability of human societies, which is the focus of this book, the pragmatic and theoretical problems are severe, entrenched, and sometimes contradictory. Indeed, there is a growing social movement centered mostly in the United States (US), the Anti-Civilizaton Movement, that harbors the goal of bringing down "civilization" because they deem the larger structure—industrialization, capitalism, and perhaps even agriculture itself—as a violent, predatory, and imperial project that devours everything in its path. This movement does not believe that "civilization" will curb its own appetite, and therefore must be forced to stop by strategic projects that will bring immediate crisis. This movement is born of individuals who believe that true sustainable social systems will never be a part of the current modern world-system, and from the startling trends and contradictions of this modern industrial society, that part of their argument is compelling. The leaders of this movement suppose that the sooner civilization collapse is initiated, the less people will suffer in the end because, if it is not stopped, it will undermine the critical life support systems of the Earth only to the benefit of a ruling global elite and create planet-wide death of people and non-humans of proportions heretofore unheard of across human history.

At the same time, there is a growing movement organized against global environmentalism that rejects that any problems exist whatsoever for the continued sustainability of modern capitalism and liberal democracy, and I have personally spent years studying this effort to better understand it. The Environmental Skepticism Countermovement and the Climate Denial Countermovement reject that phenomena such as biodiversity loss and climate change are real or important (Jacques, Dunlap, and Freeman, 2008), despite enormous scientific evidence that these trends not only are real, they threaten human wellbeing. This countermovement's central goal is to serve as a rear guard to defend against any changes to the contemporary world capitalist system and protect the world of economic production from being controlled or shifting power to the social world of deliberation and regulation (Jacques, 2014). These two movements epitomize radical perspectives at two ends of a spectrum, where forms of contemporary world society are irredeemable and must totally end, or are beyond repute causing no harm or inequality. The interesting thing is that they are both born of a similar reaction to growing sentiment that the current political-economic world-system is not sustainable. As world ecological systems increasingly are in crisis, the Anti-Civilization Movement reacts to stop these crises, and the Environmental Skepticism Countermovement reacts to defend the dominant political-economic and value systems from criticisms that emerge as a result of these crises. This book will examine the issues of sustainability from different perspectives, and there is disagreement about the specific trajectory of sustainability. Some analysts believe that current consumption, if measured in terms of non-declining welfare, is sustainable (Arrow et al., 2004). This simply means that the levels of all current welfare do not get worse. Thus, affluent societies do not become poor, and poor societies do not become poorer. From the perspective of the poor, maintaining the status quo, however, is not terribly appealing and can even be considered a type of criminal foreclosure on millions of peoples' lives. However, most of the literature in sustainability studies indicates that world civilization can neither sustain current consumption levels nor growing consumption and population. (The response to Arrow et al. is through Daly et al., 2006.)

The sociologist Urich Beck (1999) has argued that the basic forms of modernity that emerged out of the Western Enlightenment embodied in modern science, capitalism, and the nation-state are all organized around controlling nature, knowledge, and people. This control, ironically, causes crisis that is the opposite of control, and failure draws a crisis of legitimacy for capitalism, science, and the nation-state through reflexive observation by citizens around the world. Beck argues that such awareness initiates a new phase of "reflexive modernity" where individuals and groups create unexpected alliances across national borders to fight those who threaten to bring more disruption. However, what Beck did not anticipate was a countermovement to this reflexive modernity and to reflexive citizenship, and even to certain forms of science (e.g., ecological science that makes potential impact on policy versus industrial-type science geared to facilitate economic production). This anti-reflexivity, organized through the Environmental Skepticism and Climate Denial Countermovements, has proven surprisingly influential, even though it has very parochial origins—in general, it comes from exclusively a minority of conservative elites almost exclusively in the US and some in the United Kingdom (UK) and has virtually no scientific support. In some ways, we might view the battle for sustainability as a battle for reflexive thinking that is aware of the past and concerned for the future, which approaches human projects with humility and deference to a broader complex, self-organizing, dynamic but homeostatic ecological setting in which all people must live. If ecological functions are changed enough that they are no longer complex (simplified), able to self-organize (denuded, such as through desertification), or function under dynamic homeostasis (the metabolism of any system is disturbed, where steady flow of inputs and outputs like our own breathing of dynamic input and output of oxygen and carbon dioxide produce a stable respiratory system), then complex societies cannot continue to develop in these areas.

Still, thinking about sustainability is not a simple matter that we can all easily agree on, even when we are attempting to be reflexive. Consequently, a coherent approach or lens is needed to discern patterns among a lot of noise and disorder. The approach I employ here is a systems approach, explained more in Chapter 2, which means that this text assumes the projects of sustainability operate at a systems level, and that sustainability and problems of sustainability are systems problems. The benefit of this approach is that it takes a structural view of a very large problem, where we are mainly concerned about the architecture of relationships and the scaffolding that connects individual parts of the world to make up ecological and social arrangements. A systems approach provides us a theoretical approach we can use to understand the problems of sustainability. Further, the systems approach is representative of a lot of literature in sustainability because many sustainability scholars look at sustainability in this very way.

Within this systems level, there are recurring concepts worth introducing early on: **institutions**, **capital**, and **ecosystem services**. Political scientist Oran Young (2013) explains, "Institutions are collections of rights, rules, and decision-making procedures that give rise to social practices, assign roles to the participants in these practices, and guide interactions among the participants." Institutions are core features of any governing system, whether formal or informal, local or global. Related, regimes are institutional "orders" of a specific system, and this term is used both for ecological regimes and for political regimes. Political regimes are institutions for a specific issue or geographic area.

Capital is any kind of asset, and there are several forms. Human-made capital includes physical, financial, human, and social capital. There is also "the inheritance that all humans receive from nature in the form of terrestrial, oceanic, and atmospheric resources that generate flows of services, called ecosystem services," which is natural capital (Brondizio, Ostrom, and Young, 2009). Physical capital is stock of human constructed resources such as infrastructure (roads, dams), tools, and factories. Financial capital is money and can be used to procure other types of capital. Human capital is the knowledge and skill that are found in a human population, and social capital is a network of knowledge, trust, institutions, and systems of reciprocity.

Ecosystem services are natural capital, as noted, but are the ecological goods and services that provide critical life supports for human societies. Ecosystem services come in four types: provisioning, cultural, regulating, and supporting. Provisioning services are goods like timber provided by nature. Cultural services, like clean water for ceremonies or sun bathing at the beach, are goods and services that provide meaning and recreation to people and groups. Regulating services, such as a stable climate, provide stability. Supporting services, like soil formation, allow for all other ecosystem services. A series of studies performed by one of the largest assemblies of natural and social scientists in history, the Millenium Ecosystem Assessment (MEA) (2005a), has shown some direct and indirect ways ecosystem services provide critical life support systems. Figure 0.1 is an abbreviated representation of these services.

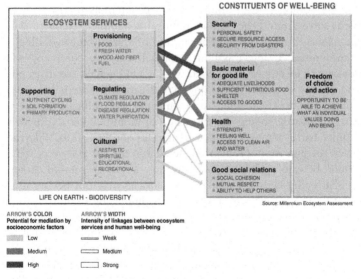

Figure 0.1 Ecosystem services and constituents of wellbeing
Source: Millennium Ecosystem Assessment

DRAWBACKS TO A SYSTEMS APPROACH

While the systems approach has important benefits, any lens we choose to view an issue we obscure some things even as we illuminate others, and the systems view has several weaknesses. One weakness is the assumption that social systems have similar dynamics as ecological systems, and, as you will read in this book, I believe this to have substantial merit at the macro, or large-scale level, of social groups over longer periods of time. However, when using this approach, one danger is to obscure important politics that differentiate people when subsuming them into an abstraction of "the system." For instance, we may be tempted to think that inasmuch as social and biophysical systems are alike, cycles of social exploitation, inequality, and misery are quite "natural" and therefore inevitable when, in fact, they are unequivocal human constructs made by human decisions that are by no means pre-determined. Indeed, if our future is out of the hands of human agency, there is little reason to study sustainability if we are only going to be a victim of it.

Because of these problems with a systems approach, I use it with some trepidation, particularly since the abstraction of systems may also lie complicit with ambivalence and attempts to merely problem-solve around the current political-economic systems to keep *the political-economic systems* going instead of vibrant human-ecological communities. Using a systems approach may be consistent with attempting to simply "hold what we got" and defend the status quo; and, to my eyes, the status quo world capitalist system is a major source of our sustainability problems and it is difficult for me to imagine that defending the status quo of infinite material growth in a finite Earth system is at all workable for another 50 to 100 years, given the enormous problems generated in the last 100 years. Thus, reader, beware of dragons that may hide in any theoretical choice.

That being said, other theoretical choices may not be any more helpful. If we were to take the conditions of dominant social assumptions and adopt a technological optimism around individuals and their rationality, we would likely come to conclusions that we can shop our way to sustainability and build robots and other advances to solve any problem that arises. While technology and innovation are essential to building sustainability, there is little indication that growing our way out of this problem is feasible, prudent, or consistent with equitable, just, or humane societies. More will be said of these concerns throughout the text, but it is clear that the lens we put on will show us some things and obscure others. The systems approach is compelling to me and perhaps to others who use this lens, because it puts some important goals in context. For example, how much biodiversity should we preserve or how much land should we deforest for agriculture? A systems approach appears to allow for some pragmatic balance—if we deforest too much, we tip the biological system into a different state that then undermines our ability to have any timber or the other wonderful life of the forest. If we don't consume enough, our own material needs will collapse.

JUSTICE

Another clarifying lens this book uses is that sustainability inherently involves dealing with problems of justice—what is fair, equitable, what is deserved, and what kind of distribution of goods and resources are normatively preferable, not only for current generations, but future ones as well? When someone is wronged, such as a poor minority community targeted with toxic waste from another area or country that causes disease, what is the good and right way of treating this situation: should the violators pay reparations? Should the ones responsible be penalized?

We might think of sustainability as a process whereby others realize their rights to life, both now and in the future. This means that decisions to disturb ecosystems may foreclose on the ability of others to benefit and realize their rights. However, do we include non-humans in this calculation, where no organism should live systematically at the expense of others? We might also conceive of justice as a series of competing loyalties to larger and larger groups we are a part of, from our families to the human species, and depending on what groups we identify with determines the form of action that is just (Rorty, 1997). In another view, political philosopher John Rawls (1971) proposed that just institutions are those that ensure that any inequality that society allows must benefit the least well off of that society, otherwise such inequality is unjust.

In either case, it is clear that egregious systems of injustice create death and disease systematically for some groups, and justice would require increasing improvements and changes to this kind of social arrangement, not stagnating or decreasing welfare.

In this text, we need not embrace any specific equation, because different societies have alternate ways to produce social justice—such as consensus decision making over Rawls' difference principle or Rorty's spheres of loyalty—but rather this book assumes that if human sustainability requires a thriving human population, there cannot be vast inequalities that lead to deprivation, misery, and death as a regularity. Anyone systematically made seriously worse off, and who was not well off to begin with, deserves redress. This is difficult because we have a plurality of nations, cultures, and countries that have competing demands, and the formation of just and sustainable institutions may be more of a work in progress across time than it is a race to the finish line; but, in a world with just and sustainable institutions, inequalities, distribution of goods and bads, and responsibilities cannot be continually made worse and more unequal. We may come to form just institutions in different ways around the world, but the process must be one of improved equalities, fair distribution of goods and bads, and deserving responsibilities. This is such a critical form of sustainability that justice is part of the basic formation of our First Principles described in Chapter 2.

The rest of the book is organized to deal with major problems and principles of sustainability, what kind of attitude is reasonable (e.g., optimism), how to measure sustainability, ethics, and moral systems that drive sustainability problems and offer promise, important political-economic conditions, and the history of collapse. However, each chapter deals with a main storyline with vocabularies and methods that help make sense of the wild and woolly notions important to sustainability studies. It should be noted that the **three Es** (Ecology, Equity, and Economics) are integrated across the book rather than having dedicated chapters to each focus, though there is a chapter on the important broader moral concerns involved in sustainability.

Chapter 1 will set the stage for thinking about sustainability, where it is made up of contradictions between growth and the integrity of systems that feed this very growth. To grow, we consume, but consuming means we exhaust resources and threaten growth at some threshold. We will also explore the use of responsible knowledge and what is at stake and what social and ecological changes bring us to the place where we even need books that think about the sustainability of world society.

Chapter 2 will explain that sustainability is an "essentially contested" term that cannot be resolved by debate. The term has a generally agreed upon but vague and ambiguous basic meaning, but when we move to put it in practice, there is no agreement on what sustainability looks like. This is especially true when negotiating the fault lines between how much ecological protection, economic growth or development, or equity is needed or right for achieving sustainability. However, there is a recurring problem structure for sustainability found in the general contradiction discussed in Chapter 1; and, there is a remarkably consistent agreement about general principles of sustainability. From this consistency in sustainability studies, this chapter summarizes the First Principles of sustainability that are important for the rest of the discussion throughout the book. These First Principles explain critical ecological life support systems must remain intact for societies to last, and that, then, these societies must develop restraint and justice. The chapter also explains basic system dynamics and the Adaptive Cycle that appear relevant for many social and ecological dynamics, and that provide us the framework for integrated social-ecological systems.

Chapter 3 retells recurring storylines important to sustainability, having to do with population and resource scarcity concerns, the role of optimism and pessimism for solving problems or denying them, and in this chapter we tackle the enormous issue of sustainable development.

Chapter 4 discusses the various ways that researchers have attempted to measure sustainability. These measures include measures of consumption, such as the popular "Ecological Footprint" approach as well as other less straightforward ways. Other approaches include the well-known models from *The Limits to Growth*, and heuristics like Planetary Boundaries and the so-called Triple Bottom Line model. While each method differs substantially in its approach, they all provide a consistent and sober warning about the future.

Chapter 5 deals with the ethics that any sustainability calculation or discussion must confront. In some ways ethics helps explain why some social-ecological problems are produced, as well as a way to think of core requirements for sustainability in the future for the second tier requirements of First Principles that cover prudence, restraint, and justice.

Chapter 6 explains important political conditions for sustainability that require robust institutions to solve the "tragedy of the commons" and other collective action problems that are in play from local to global governance. We also track the direction of global environmental governance, and key challenges to future global environmental politics, including the challenges for civil society and social movements.

Chapter 7 works through the complex causes of civilization collapses in the past, as well as the nature of social crises that follow collapse, Dark Ages. We look at the latest models of collapse from the best science and examine the collapse of the Classic Lowland Maya as a case. Important lessons can be found in these models and cases because the basic requirements for sustainability corroborate the First Principles explained in Chapter 2.

The conclusion of this book attempts to reflect on what these concerns synthetically mean for sustainability now and in the future.

SUSTAINABILITY
WHAT IS AT STAKE?

MAP OF THIS CHAPTER

The central storyline of this chapter is that of contradiction. We first focus on the central contradiction between growth and integrity, where using more resources undermines the very relationships that are required for continued growth. Growth also has a distributive effect on life chances for others and fundamentally involves justice problems. This essential contradiction is evident in the collapse of the Western Roman Empire described here first. The chapter will highlight a number of concepts dealing with the era of human domination, the Anthropocene, and the opportunities and challenges this presents. The chapter also proposes that we are obligated to use knowledge responsibly and discusses the problems involved in this pursuit. The chapter then discusses the context, extent, and meaning of these global environmental and social changes, and some of the particular threats these changes pose to people, that make studying sustainability necessary.

WHAT LIES BEYOND THE GATES OF ROME?

At midnight on August 24, 410 CE, Roman slaves slipped open the Salarian gate to the city and the Goths famously sacked Rome.

This was the beginning of the end of one of the most powerful and influential empires in history, and "By 476, Rome was the fiefdom of Odoacer, king of the Goths" (Ferguson, 2010). This collapse is also one of the great mysteries that has generated thousands of volumes of work, analyzing how and why the Roman Empire fell. Like issues in sustainability, there are competing ideas about how to look at the problem. What we take away and remember beyond the gates of Rome and other historical collapses is essential for future adaptation and resilience of human society.

Ancient Rome was one of the cornerstones of Western society. Romans built enormous architectural and engineering monuments, its social structure was highly complex, and its political organization was legendary. Our calendar, many military strategies, and many contemporary Western legal traditions date back to this period, amongst other important legacies. Rome was an empire and therefore its center of power ruled over many other areas. Rome controlled vast areas of Europe, stretching across western Asia and northern Africa, which it used as a granary and tax base. Early on, Rome conquered these areas to solve resource shortages. However, the imperial approach produces a contradiction: As expansion and conquest fed the imperial center of power, new Roman territories required more resources and complicated the social dynamics for governing. More conquest produced more needs and, over time, the Empire hit a critical moment where these needs could not be met adequately. The classic scholar of Rome, Edward Gibbon (1994) wrote:

> the decline of Rome was the natural and inevitable effect of immoderate greatness. Prosperity ripened the principle of decay; the causes of destruction multiplied with the extent of conquest; and, as soon as time or accident had removed the artificial supports, the stupendous fabric yielded to the pressure of its own weight.

Scholarship of the Roman collapse has changed substantially since Gibbon originally wrote these words in 1781. For example, Jones (1964) wrote that over-taxation undermined the Roman economy, leaving peasant agriculturalists without enough to survive, setting up the Empire for collapse. This view has been recast though, with evidence that peasants were not being over-taxed and the economic productivity from agriculture was booming in the fourth century, "with no sign of overall population decline" (Heather, 2006). However, the Empire had maximized its agriculture productivity and could not increase it substantially and, at the same time, barbarian tribes had evolved to respond to Rome and had allied together to create a very large threat. Because Rome had grown so much, it had generated this antagonism and opposition; and, because it had maximized grain output, it could not raise a larger army to fight this opposition.

> The west Roman state fell not because of the weight of its own "stupendous fabric", but because its Germanic neighbors had responded to its power in ways that Romans could never have foreseen. There is in all this a pleasing denouement. By virtue of its unbounded aggression, Roman imperialism was ultimately responsible for its own destruction.
>
> (Heather, 2006)

Even as the Roman project lasted more than 1000 years (the Empire lasted more than 400 years) and had withstood attacks and challenges for much of this period, it became increasingly vulnerable to economic and social instability that were connected to the food supply and the use of Roman power to deliver it. Eventually, critical events cast the Empire into collapse during the fifth century CE. The Western Roman Empire was pounded by Germanic invaders in 406, and sacked in 410, noted above. The Huns stretched Roman resources to the brink of exhaustion. By the middle of the fifth century, this exhaustion led to the loss of control over England, much of its European base, and—critically—North Africa. The Vandals took control of African territories one by one, removing grain supply and revenue that had been essential to keeping Rome from teetering into the abyss of oblivion.

With these losses, Rome became a classic example of social collapse. Such collapse has occurred with regularity across human history and often with breathtaking and shocking velocity. Societies can appear stable, just as a forest appears stable before a forest fire, but then at unpredictable times, small changes can hurl the society into chaos and disintegration. Causes for these social collapses are typically overlapping economic, social, and ecological problems. This book will consider some of the lessons from sustainability studies and science at a time in human history when we, as a species, have begun to radically change core ecosystems and cycles. Some of these lessons are quite dire, but we have advantages that Rome did not have, obviously including a good understanding of Rome's own demise and a good understanding of other civilization collapses. Of course, the record of people learning from history is notoriously spotty, and anthropologist Patrick Kirch (2005) opines:

> Yet dare we hope that such retrospective understanding of how humans have transformed the Earth—and in the process suffered through a panoply of crises, social collapses, and restructurings—could possibly be of use in guiding our collective future? Some at least think that the archaeological record provides lessons that could guide our future. Whether we heed them is up to us.

Civilization collapse occurs when critical needs are not met for that civilization and it is what happens when a civilization fails to be socially, ecologically, or economically sustainable. Collapse can also result from imperial aggression and dispossession, where the target society may be subject to disease, slavery, or loss of resources, and then the empire itself requires more and more resources that are difficult to ensure. Collapse has occurred across history numerous times, and there is little reason to think that the danger of a civilization collapse has passed. In fact, many scholars in the area of sustainability warn of the threats of civilization collapse by the end of this century. Celebrated sustainability intellectual, David Orr, warns of the oncoming deep crisis that faces world civilization:

> The "perfect storm" ahead, in short, is caused by the convergence of steadily worsening climate change; spreading ecological disorder (e.g., deforestation, soil loss, water shortages, species loss, ocean acidification); population growth; unfair distribution of costs, risks, and benefits of economic growth; national and ethnic tensions; and political incapacity.
> (Orr, 2012)

The story of Rome tells us there are inherent contradictions to growth and consumption of land, resources, and social capacity to solve problems. On the one hand, expansion of Rome meant that it became more powerful, and its people enjoyed some prosperity but, on the other, this prosperity had to come from somewhere and even as the empire grew, it weakened and destabilized the social and natural sources of this growth. The contradictions of this growth set the stage for Rome's demise, and Rome is only one demonstration of the power of these contradictions. As great as any one people may believe themselves to be, they are always reliant on finite social and ecological systems. Past a certain threshold, growth contradicts the integrity of these critical systems. This is a central lesson of sustainability, and the irony is that every great society probably has known and at the same time denied this fundamental condition of history.

WHAT IS SUSTAINABILITY?

We begin here with some of the basics of **sustainability**, though these questions are explored more in-depth in Chapter 2. There are literally hundreds of attempts to define sustainability, but many of these efforts agree around a few basic broad, if ambiguous, values. My reading of these literatures is that sustainability is the imperfect process of building and maintaining global social systems of capable, accountable, adaptive, just, and free people who can make important decisions and trade-offs with foresight and prudence and who foster the robust, self-organizing, dynamic, and complex ecosystems around the world for now and future generations.

In the last few decades, sustainability studies and science has grown around an increasing concern that the modern, interconnected global economy and population is moving far away from the above aspirations and is pushing Earth systems to their limits where they will no longer be able to support the human prospect in the same way.

The Latin root of the word "sustainability" is *sus tenere*—to "hold up" or "maintain," perhaps just as Atlas is said to hold up the heavens in Greek myth. However, the most widely used definition of sustainability comes from the World Commission on Environment and Development (WCED), otherwise known as the Brundtland Commission. WCED (1987) defined sustainability as development that, "meets the needs of the present without compromising the ability of future generations to meet their own needs." But these two definitions are really just the beginning, because global sustainability involves complicated questions about economic growth, ecological integrity, and justice around the world.

The objectives of this book are to explain the central debates and advances in the field of sustainability studies, and maintaining *global* social and ecological systems. Buckminster Fuller captured the sentiment of planetary sustainability this way: "*making the world work for 100% of humanity, in the shortest possible time, through spontaneous cooperation, without ecological offense or the disadvantage of anyone*" (Fuller, 2008, emphasis added). The vocabulary to refer to 100 percent of humanity is awkward, where we might refer to world civilization, world society, global societies, but the central concern is to think about a holistic, planetary sustainability.

If global welfare is a worthy goal of sustainability, then we are left to reconcile how growth and the accumulation of wealth are arranged with limited Earth systems to feed and deal with the waste of growth. This means that as Earth systems are exhausted, some present groups and future generations may have less, across local, regional, and global contexts, and this growth then has a distributive effect on wealth and welfare. Since growth and its waste affects and trades off current and future chances for people to live well, sustainability is a justice problem.

CHECKPOINT: RESPONSIBILITY AND KNOWLEDGE

Why do we need to discuss sustainability? Would it have been useful for Roman senators or citizens to think about problems of sustainability? If they did, what claims would have merited attention at the time, and if there were warnings of a coming crisis, how responsible would the leaders and emperor of Rome have been to following through on such alarms?

Are there really problems that threaten human societies that are *that* serious? And, really—how do we know? Perhaps the problems of sustainability are just what social critic H. L. Menken warned about when he said that "The whole aim of practical politics is to keep the populace alarmed (and hence clamorous to be led to safety) by menacing it with an endless series of hobgoblins, all of them imaginary" (Holloway and Sou, 2002). We should take Menken's warning seriously, which means we need to have a good answer to "how do we know" that there are serious sustainability problems that are not just meant to scare populations into submission to a state, corporation, or faction.

When it comes to global environmental changes, such as climate change, it is important for us to ask: "How do we know what we think we know?" The study of how we know what we think we might know comes from the philosophical tradition of **epistemology**. There are a lot of ways that individuals come to believe something is true, and the veracity, or truth, of that knowledge is permanently uncertain.

Consequently, we need a strategy to evaluate assertions. The strategy in this book is to rely primarily on technical scientific literature to describe environmental problems because outside of this literature claims receive less oversight and scrutiny. This does not mean orthodox science is always right, but there is a system of correcting what it gets wrong over time. If we rely on assertions outside of the scientific community, for *scientific* questions, we can easily be caught in the crossfire of ideological conflicts without even knowing it.

Environmental groups may often report problems with accuracy and diligence, but they are political groups and we must expect them to cherry-pick their evidence to promote their goals. We must expect the same from industry and governments, who often serve political and economic elites. Often, government and international governmental organizations provide accurate information, but, they too, have important political agendas and we should use these sources consciously and with purpose. The scientific literature is imperfect and it has its own biases and politics, but it has systems in place that guard against overt propaganda.

The epistemology used in this book assumes that all knowledge is political, and therefore is never an un-slanted vision of reality—all knowledge is produced, distributed, and sustained with a purpose. On the other hand, the epistemology used in this book also assumes that the human condition is not so subjective that reliable knowledge is impossible. One strategy to negotiate the forces of political subjectivity and establish a defensible understanding of the world is to corroborate points of view over time. When some form of reality can be described similarly by multiple people who have different interests and perspectives, then that knowledge is more reliable.

The process of oversight and corroboration in science comes from peer-review, which provides a system of good faith witnesses. Experts weigh-in on every publication, including this book you are reading. Authors must respond to criticism from others who are experts in the field and revise their work accordingly. Over time, corroboration openly builds confidence and the body of knowledge grows.

This approach is consistent with Norton's "limited realism" that admits that we are forced to individually interpret the world, but experience and time winnow mistakes. Our belief systems and behavior that come from our understanding of the world, "must stand the test of more and more experience on the part of more and more observers and their inputs over time" (Norton, 2005). Corroboration by multiple voices provides a shared judgment to build common ground for sensible action. Indeed, the ecological thinker Aldo Leopold warned that human survival will itself test the validity of our beliefs and decisions.

That said, there is rarely unanimity in communities of scientists, and responsible use of science indicates we work with the consensus that is forged in these communities, even with the understanding that it is always possible for minority dissidents to be right in the end.

This means that it is irresponsible to ignore scientific findings about the human condition just because we might not like what these findings say or mean to us. David Orr (2002) has written that educating ourselves about changing environmental conditions is one of the central problems for sustainability. Developing a more creative citizenship that takes responsibility for these changes is another central challenge. This book is written with the assumption that the orthodox science actively maintains various political agendas and its conclusions could be quite wrong, but we have a responsibility to attend to claims that have the most scrutiny and corroboration over time.

So, what does some of this literature indicate? For one thing, what the scientific literature says and what the public often thinks it says are disconnected. The clearest example of this issue is climate change. While there is public debate, especially in the United States (US), about the reality of human-caused climate change, the basic science of global warming has been in place for a long time and has not been disproven. There is a clear and robust consensus in the scientific community that the Earth's climate is warming, and that human emissions are the dominant cause. But, this is just the beginning.

GLOBAL ENVIRONMENTAL AND SOCIAL CHANGE

Contradictions that existed before the fall of the Roman Empire may have contemporary corollaries for a growing world economy, population, and consumer culture. Things have changed so very quickly for us and environments we live in, particularly in the last 50–100 years; compared with our history on Earth these changes should be quite startling.

All organisms that exist on the Earth with us now are a result of a complex and beautiful dance that has worked for eons. *Homo sapiens* evolved within the great ape family, where our last common ancestor to this family was 5–6 million years ago, but modern humans have only been around for about 200,000 years. Like other species, we have evolved under specific conditions and needs that are fulfilled by the planet's provisions and services that come from specific ecological systems and cycles, **ecosystem services**, as well as the remarkable human ability to think abstractly, cooperate with each other, and to plan ahead.

Our ability to modify our environment to our own needs is truly astounding and unprecedented in Earth's history. This ability has led to revolutionary advances in how people live: through the **Agricultural Revolution** that allowed sedentary civilizations some 10,000 years ago, the **Industrial Revolution** some 250 years ago, and the **Information Revolution** only decades ago that has produced computers and the Internet. From these advances, we have tools like modern medicine that allow for a much longer lifespan, at least for people who have access to these advances. Our capability to grow more food on the same amount of land has dramatically risen over the last 50 years, and this has fed a population that has grown to more than 7 billion people, a demographic change that added 6 billion people in only 200 years.

Thus, agricultural productivity kept up with giant leaps in population during the twentieth century. However, global food production needs to grow another 70 percent (and 100 percent in poor countries) of what it was at the beginning of the twenty-first century to keep up—"The number of people which the world must feed is expected to increase by another 50% during the first half of the century," not to mention increasing nutrition for the poorest of the poor in the world (Hertel, 2011).

Our ability to change the Earth, just like the contradictions of growth above, cuts two ways. We have succeeded in building a well-populated species within a global economy, but this growth has engendered critical contradictions. At the same time that food availability needs to grow enormously in a few decades, growth in critical crop yields are slowing, soils are degrading, biomass is being transitioned for energy (ethanol) instead of food, water availability is diminishing in many areas of the world, key inputs—like non-renewable phosphorous fertilizer—may be depleted between 2050–2100, while climate change is expected to reduce food productivity "precisely where malnutrition is most prevalent" (Hertel, 2011). These changes may create "perfect storms" of agricultural losses in localized areas (Hertel, 2011), and dramatically affect prices of land and food, all of which may cause social crises, particularly in urban areas in poor countries.

Thus, human capacity has undercut ecological systems that are the wellspring of our development. This is the contradiction of growth and progress central to debates of progress and collapse found in sustainability—just as the growth and progress of the Roman Empire planted seeds for its own demise, and "Prosperity ripened the principle of decay" (Gibbon, 1994).

In the history of life on Earth, we discover that all species eventually go extinct. Clearly, the human species must face the fact that we are not exempt from this universality. Consequently, we have something like a "lifespan" that thus far has only been around 100,000–200,000 years.

Measured against the 300 million year lifespan for some animals like trilobites, our current tenure is only a planetary instant. However, humanity has initiated enormous changes at the planetary level, especially measured against our own brief existence. Humans have only lived with sedentary agriculture for the last 10,000 years, and industrial production for only the last 200 years. Yet, these latter two periods forced accelerated global environmental changes to primordial systems that took eons to develop, with intensifying impacts after World War II. This means that humans have made huge changes to ecological systems in only a fraction of their time on Earth—very quickly.

Interestingly, analyses of skeletal remains indicate that general human health was best before the Agricultural Revolution, and after this change, average lifespan actually decreased (Angel, 1984). It was not until the twentieth century that (affluent countries) saw lifespans move past 40 years, and now the average lifespan for people in the United States is more than 70 years for men and women (Wells, 2010).

More than half of the ice-free land has been altered by people for agriculture, roads, infrastructure, and settlements, removing habitat for other organisms and interrupting critical ecological services to people. Of all the solar energy converted by plants across the whole planet every year, between 37 and 50 percent is taken by people, decreasing the supply of energy for the other 5–15 million species and this is in itself a threat to human sustainability. And, while humanity has improved our ability to survive amongst the elements, it is a general principle of evolutionary theory that as a population increases in a single habitat, at some threshold, the chances of survival decline. To what extent will our ability to modify the environment defend us from evolutionary pressures, and to what extent can we change the Earth systems across habitats until our survival chances decrease? How much will political leaders pay attention to this balance?

Indeed, already humanity controls about half of the global available freshwater, and we have changed the water cycle by altering rivers, converting wetlands, and changing the way water is distributed around the world, facilitating profound shifts in larger food webs. Ironically, as we have exerted more control over water cycles, *higher* portions of the human population have lived with water shortage— in 1900, only about 2 percent of the world population lived with water shortage, in 1960 it was 9 percent, and by 2005 as much as 35 percent of the world lived with water shortage (Kummu, Ward, de Moel, and Varis, 2010). By 2050, the number of people living with water shortage, in part as a result of climate change, is expected to rise tenfold (McDonald et al., 2011). These are clear contradictions to the wellbeing and sustainability of the human population.

Lost and fragmented habitat, in conjunction with other factors like pollution, has produced what is called the **Sixth Great Extinction** of life on Earth. All species go extinct, but tend to exist for 1–10 million years, which means there is a normal rate of extinctions, which we know from fossil records that span hundreds of millions of years. However, there have been five punctuated extinctions during the Ordovician, Devonian, Permian/Triassic, and Cretaceus/Tertiary periods. Biologists explain that the Sixth Extinction is evident because extinctions of flora and fauna are at least 100–1000 times the normal extinction rate.

In this sense, we do not need to wait for a collapse, it is already occurring in the non-human world. **Biodiversity** loss has and will continue to alter the future evolution of life on Earth, reducing the gene pool, making life more homogenous, and fostering species that flourish in human modified environments (e.g., vultures, squirrels, and coyotes), including invasive species, pests, and weeds. "By contrast, the current extinction resembles none of the earlier ones and may end up being the greatest of all" (Şengör, Atayman, and Özeren, 2008).

And, biodiversity loss is more than a loss of treasured Earth companions. When plants and animals go extinct, key functions like pollination are lost, threatening the sustainability of ecosystems around the world because biodiversity tends to stabilize ecosystems, and loss of biodiversity causes major instability in ecosystems. Consequently, biodiversity loss threatens human sustainability:

> Changes in diversity can directly reduce sources of food, fuel, structural materials, medicinals or genetic resources. These changes can also alter the abundance of other species that control ecosystem processes, leading to further changes in community composition and vulnerability to invasion.
>
> (Chapin et al., 2000)

Among this biodiversity loss is the loss of crop varieties that are cultivated by different cultures. Indeed, the richest areas of language and ethnicities correlate with areas of both flora and fauna diversity. There is an incontrovertible link between plants, animals, and lands that people benefit from, and the knowledge systems, linguistic development, and cultural identity that grows alongside these ecological niches. Both are being lost through powerful forces that are:

> placing the world's diversity in both nature and culture increasingly at risk. *This means no less than placing at risk the very basis of life on Earth as we know it*: the natural life-supporting systems that have evolved on the planet, and their cultural counterparts have dynamically coevolved with them since the appearance of *Homo sapiens*.
>
> (Maffi, 2006, emphasis added)

Areas of rich biodiversity and cultural diversity show a "parallel extinction risk" (Schäfer, 2012), which means species and cultures are being extinguished simultaneously, in part caused by "dramatic loss of livestock breeds and agricultural varieties as well as traditions for raising them, and erosion or obliteration of regional cuisines and foodways," and, as diversity is being lost to homogeneity "almost everywhere" "forces promoting homogeneity are playing an endgame on a global scale" (Redford and Brosius, 2006). "Endgame" is another way of saying total ruin and collapse.

Industrial society has also changed the way carbon moves between land, air, and water. Carbon naturally is mobilized from the land, for example during forest fires or decomposition of plant and animal life, into the atmosphere and the oceans and back into the air and land again, and this is called the **carbon cycle**. Prior to the Industrial Revolution, the human contribution to the carbon cycle was trivial given our small populations and smaller changes to the land. However, after the Industrial Revolution, burning coal and oil became the foundation of energy for industrial societies and emissions of carbon dioxide (CO_2) have dramatically risen as a result. As of this writing, atmospheric CO_2 has increased by ~40 percent, compared with pre-industrial levels.

CO_2 is a gas that absorbs heat in the atmosphere and human emissions of greenhouse gases is the dominant reason for climate change evident since the twentieth century. Increased global average temperatures by 0.85 [0.65 to 1.06] °C, and "Warming of the climate system is unequivocal, and since the 1950s, many of the observed changes are unprecedented over decades to millennia" (IPCC, 2013). Of this warming, 90 percent has been absorbed by the **World Ocean**, and if this heat were instantaneously transferred to the atmosphere it would increase the Earth's temperature by an average of 36°C (65°F) (Levitus et al., 2012). Warming of the ocean changes the way the ocean water moves around the world, because ocean currents stabilize Earth's climate. Changing the marine system this way also threatens marine organisms that live there and tinkers with the entire Earth's water cycle. Global warming has had a multitude of confirmed impacts—melting glaciers, sea level rise, overall biodiversity loss, pest outbreaks, rapid changes to the marine food chain in the ocean, droughts, floods, and crop changes (that in some areas improves, in some areas declines), and a large reduction in summer sea ice in the Arctic. In addition, CO_2 is mainly absorbed into the ocean from the atmosphere, and if there is too much, carbonic acid acidifies the water. Ocean acidification threatens fish, coral reefs, and any organism that has a body that includes calcium carbonate, such as the organisms that build the shells we find on the beach.

Indeed, biodiversity loss may be most profound in the ocean. Fishing, habitat degradation, pollution, species invasions, climate changes, acidification, and the run-off of nutrients (nitrogen and phosphorus) are reducing biodiversity and function of the World Ocean. Marine scientist Jeremy Jackson writes in the *Proceedings of the National Academy of Sciences*:

> Today, the synergistic effects of human impacts are laying the groundwork for a comparably great Anthropocene mass extinction in the oceans with unknown ecological and evolutionary consequences. Synergistic effects of habitat destruction, overfishing, introduced species, warming, acidification, toxins, and massive runoff of nutrients are transforming once complex ecosystems like coral reefs and kelp forests into monotonous level bottoms, transforming clear and productive coastal seas into anoxic [low in oxygen] dead zones, and transforming complex food webs topped by big animals into simplified, microbially dominated ecosystems with boom and bust cycles of toxic dinoflagellate blooms, jellyfish, and disease.

> (Jackson, 2008)

The "**Anthropocene**" is the term used to describe the current period where humans dominate all ecosystems, and Jackson indicates we are stripping down complex food webs and replacing them with simple, impoverished, and denuded marine systems.

These **structural ecological changes**, or changes to the basic organization of Earth systems and cycles, have led a broad array of scientists, such as physicist Stephen Hawking, who doubt that humans will live another 100 years (in Höhler, 2010), to many other thinkers and philosophers to speculate that humans pose the greatest threat to ourselves (Western, 2001; Fuller, 2008). Others warn that we have violated the boundaries for a comfortable living space for humanity (Rockström et al., 2009). Some scientists believe that the increased interest in sustainability comes from, "evidence that humankind is jeopardizing its own longer term interests by living beyond Earth's means" (McMichael, Butler, and Folke, 2003).

Still others, outside the scientific literature, have argued that these changes are mythological or over-exaggerated, and that, no matter what, people are smart and will adapt to any problem with economic ingenuity and technological advances. This cornucopian **environmental skepticism**, or **environmental denial**, is the position that there are no authentic global environmental problems that threaten sustainability. While this position is an outlier scientifically and ethically, the perspective is consistent with the dominant Western cultural belief that maintains a strong faith in the efficacy of science and technology, and in future abundance, regardless of the limits to Earth's actual systems, cycles, and resources.

However, we can see the effects of global environmental changes already in increased disease and mortality; these environmental changes affect the availability and strength of pathogens, the availability of food and water, and other critical health dynamics. In one assessment published in *The Lancet*, authors warn of "health effects due to the social, economic, and political disruptions of climate change, including effects on regional food yields and water supplies" (McMichael, Woodruff, and Hales, 2006) that are unfortunately hard to quantify directly. Further, these changes to critical Earth systems and cycles appear to provide a substantial obstacle to further development of the poorest countries.

Perhaps the most troubling aspect of all these changes is that they are "**synergistic**" as Jackson writes above. Threats often converge to cause other problems that feed into yet other problems. Scientists agree that:

Multiple environmental stresses produce more than additive effects. They create synergies through interaction and produce quantitatively and qualitatively different outcomes from single factors acting alone. Such outcomes are derived from nonlinear processes operating on multiple spatiotemporal scales, and these lead to critical thresholds or points at which either rates of change shift dramatically and/or the system shifts into a different state. However, many, if not most, nonlinearities are unknown, and gaps in understanding these phenomena lead to gaps in knowing how to respond to them in terms of design of policy and management approaches.

(Miles, 2009)

In sum, human activities, especially since the Industrial Revolution and accelerating after World War II, have altered basic operating conditions for life on Earth by altering the chemistry of soils, atmosphere, and fresh and marine water systems, extinguishing species, and disrupting whole landscapes all in a very short time. Sustainability scholars at the highest level warn that these systemic changes to critical life support systems, in conjunction with social systems that have little flexibility, threaten *today's* global human community. Witness this assessment in the *Proceedings of the National Academy of Sciences of the United States*:

> Today, we face a set of interconnected crises that threaten the sustainability of our increasingly brittle global socio-ecological system. These include climate change, the imminent peak and decline in key nonrenewable energy resources, and a loss of biological diversity that may reduce the resilience of our global ecosystem and its ability to provide for human needs. Although most societies that declined in the past were replaced by new ones, those societies were relatively isolated, lacking the interdependency of our current global community and the interconnectedness of the crises that we face today. The possibility that our global society may suffer decline makes this a "no-analog" period in human history in which massive social or environmental failure in one region can threaten the entire system.
>
> (Beddoe et al., 2009)

ENDPOINTS

WHAT DO WE KNOW?

We know that humans have caused large-scale change to critical ecological life support systems while we have also experienced profound social and demographic changes in the last 10,000 years, but especially in the last 200 years. Changes after World War II came with longer lifespans, growth in the development of the world's economy (especially for Western industrialized countries), and an improved quality of life for many people. We also know that this growth has caused severe contradictions for human sustainability. If the Roman Empire is any comparison, as we use our capacity to conquer more and more of the world we may be planting the seeds of our own demise. Fortunately for us, unlike past societies, we have a rigorous accounting of their own collapses in the anthropological record. However, in order for that to be of value, we need to learn and act on these lessons.

CRITICAL CONSIDERATIONS

What do you think are the best ways of gaining knowledge about the Earth and its environmental systems?

What responsibilities do individuals, groups, countries, and global communities have to the body of knowledge about ecological change?

Synergistic problems from global environmental change present a complex set of problems—how do you think you would address such problems?

If contradictions of sustainability indicate that one kind of welfare may come from other sources of welfare, what criteria should be used to decide between trade-offs between economic, social, and ecological sources of welfare?

What do you think the major causes for global environmental change are?

How do you think different people in other countries and communities will feel the effects of things like changing conditions for food production or water availability?

How do you think geography affects the way global environmental changes are felt?

What are the first problems we need to address?

WHAT DO YOU THINK OF THESE SUSTAINABILITY SOLUTIONS?

1 Sustainability problems and ideas frame general education from the age of 4 years in every country. Students need to learn about their specific bioregion and the changes it has experienced, alongside ways to keep more detrimental changes from occurring. This would not be another subject of study, but the framework for all other topics. Math and science, civics and literature, could all be organized around ideas of sustainability.

2 Governments begin to think of sustainability with the same urgency as national survival and security, and put funding,

infrastructure, and policies in place to reflect "sustainability as security."

3 Governments see sustainability as a basis for social welfare, instead of a national security, and begin to put social policies in place to reflect this priority.

4 Corporations are granted certain public allowances. In return for these allowances, corporations must show how they either further or do not harm sustainability efforts. Those that are most compatible with sustainability concerns receive tax benefits, but those who create sustainability problems (such as oil companies or mining companies) must pay higher taxes.

A syllogism is an ancient form of logic where a conclusion is inferred from two or more major propositions.

WHAT DO YOU THINK OF THE FOLLOWING SYLLOGISM?

Premise A: Earth cycles and systems provide critical support for human civilization.

Premise B: The activities of human civilization are undermining Earth cycles and systems.

Conclusion: The activities of human civilization are unsustainable.

FURTHER READING

Vitousek, Peter M., Harold A. Mooney, Jane Lubchenco, and Jerry M. Melillo. (1997). "Human domination of Earth's ecosystems." *Science* 277: 494–99. In this classic summary of major structural changes to Earth systems and cycles, such as the Sixth Great Extinction and climate change, the researchers show that every ecosystem on Earth is now dominated, by which they mean deeply influenced and controlled, by human activity.

Hooper, David U., E. Carol Adair, Bradley J. Cardinale, Jarrett E. K. Byrnes, Bruce A. Hungate, Kristin L. Matulich, Andrew Gonzalez, J. Emmett Duffy, Lars Gamfeldt and Mary I. O'Connor.

(2012). "A global synthesis reveals biodiversity loss as a major driver of ecosystem change." *Nature* 486(7401): 105–8. This study indicates that "extinctions are altering key processes important to the productivity and sustainability of Earth's ecosystems" where the loss of species is having a measurable effect on key Earth system functions, at the same time and to similar degree as other global forces like ocean acidification and climate change.

Rosenzweig, Cynthia D., David Karoly, Marta Vicarelli, Peter Neofotis, Qigang Wu, Gino Casassa, Annette Menzel, Terry L. Root, Nicole Estrella, Bernard Seguin, Piotr Tryjanowski, Chunzhen Liu, Samuel Rawlins, and Anton Imeson. (2008). "Attributing physical and biological impacts to anthropogenic climate change." *Nature* 453(7193): 353–57. This breakthrough article studied biological (e.g., plant flowering) and physical (e.g., glaciers) systems all over the world and found these systems have been changing at the continental scale in a way that cannot be explained without adding in human greenhouse gas emissions forcing global warming.

Orr, David W. (2004). *Earth in Mind: On Education, Environment, and the Human Prospect.* Washington, DC: Island Press. Orr provides a sobering, yet inspiring, perspective on our current challenges and how to think about them. Orr is a global leader in sustainability thinking, and this is one of his most well-known works. In particular, he elaborates on global environmental changes as a problem of education and knowledge, and ultimately how these forces relate to human identity.

Daly, Herman and John Cobb. (1989). *For the Common Good: Redirecting Economy Toward Community, the Environment, and a Sustainable Future.* Boston, MA: Beacon Press. Daly and Cobb elegantly place key problems of economic growth into elaborate detail. This work of Daly, an economist who has argued for a steady-state economy (not based on growth), and Cobb, a theologian, contextualizes industrial economic growth as a central driver for ecological crisis in a way that continues to be relevant. This book was identified by University of Cambridge's Programme for Sustainability Leadership and Greenleaf Publishing as one of the Top 50 Sustainability Books, and it was winner of the Grawemeyer Award for Ideas Improving World Order 1992. Truly a classic in sustainability thinking.

Thiele, Leslie Paul (2011). *Indra's Net and the Midas Touch: Living Sustainably in a Connected World.* Cambridge, MA: MIT Press.

Wonderfully written, this book describes the "first law of human ecology" as we can never just "do one thing" because the nature of the world is founded on interdependence. This guiding principle leads Thiele to explore across disciplines what it means to live prudently and wisely in a profoundly interdependent world.

THE PRINCIPLES OF SUSTAINABILITY

MAP OF THIS CHAPTER

The storyline for this chapter is that of complex inter-dependence, which occurs in systems and is a defining feature of the world we live in with other people and eco-logical systems. This chapter builds on Chapter 1 by ela-borating on the recurring problem structure of sustainability, which is the essential contradiction of consuming pro-ductivity *within systems*. Because we live in a profoundly interrelated world, everything we do affects other parts of the systems humans and non-humans live in, and the systems we live in are therefore complex. The problem structure of sustainability in complex systems is exemplified in the case of the Newfoundland cod story. We will also address the fact that the idea of sustainability is "essentially contested," where there is consensus on the abstract ideas, but when we put these ideas into practice, there is less agreement about how to make economic, social, and ecological trade-offs. That said, because there is agreement on principles, the chapter ends by delineating the First Principles, or parameters, of a sustainable society.

FISHING FOR COD AND CATCHING HELL

Since the colonists landed on the shores of North America, there have been accounts of an abundant cod fishery in the northeast. Off the Atlantic coasts of the United States and Canada, an area called the Georges Bank and the Gulf of Maine fostered cod fishing for more than 400 years. By some early accounts, there were so many fish, you could get to shore from a ship by walking on their backs in the water. This was one of the richest fisheries in the world. As a result, the region's first major industry was fishing, promoting direct and indirect jobs and income that shaped the area into what it is today.

As early as the 1970s, scientists in both countries began to warn against over-fishing the cod. In Newfoundland, Canada, and in New England, USA, this was a delicate message for fishers who could trace cod fishing through multiple generations, and whom were entirely dependent on the cod catch. If the cod were being over-fished, this implied that fishing needed to be reduced, and revenue of the fishers would decline, boat and mortgage payments and savings accounts would be endangered.

On the other hand, if it were true that the cod were being over-fished and no reductions in fishing occurred, or fishing was not reduced enough, the cod themselves would be endangered, and then the fishers' entire future would be endangered alongside the fish. The contradiction of consuming productivity finds its way into this controversy, and, indeed, there were proposals and plans, protests and policy debates.

Around these debates were "how much" to cut back fishing and by what measure because it was not easy to determine exactly how many fish there were in the population. To paraphrase a well-known truism in fishing management: Counting fish is like counting trees, except you can't see them, and they move. So, fish populations are assessed by sampling the water, and even though some of the best science at the time was available to the cod fisheries, fishers and many in the community did not trust the fish scientists or the government in the assessments of how many fish were there. In the end, however, fishing was not reduced enough, and the cod fishery of the entire region suffered a now legendary collapse familiar to fishery experts. The contradiction worked something like this: hundreds of years of fishing worked as an increasing disturbance to the cod population. Indeed, early colonists believed that there was a near inexhaustible supply of fish. However, after a long time fishing the cod and later on in the twentieth century attempting to maximize the cod catch with the most powerful vessels with the most advanced technology and gear, the cod population began to show signs of weakening and not being able to survive this disturbance. In the 1990s, the population of cod crossed a threshold, or a breaking point, and the cod suddenly and catastrophically collapsed in near totality.

In Newfoundland, hundreds of years of fishing meant that almost all the area's economy was either directly involved in fishing or served the fishing industry; but, without the fish, the economy of the entire area *also* collapsed, losing between 35–50,000 jobs. Richard Cashin, Chairman of a Canadian task force to report on the challenge, put the problem this way in 1993:

> Far more is at stake than the closure of single-industry towns. The society itself is at peril. We have a tragedy of enormous proportions for the people who operated the boats, the people who worked in the plants, and for many processing and fishing enterprises, large and small, where people have laboured for so long ... . Failure of the [groundfish] resource means a calamity that threatens the existence of many of these communities throughout Canada's Atlantic coast, and the collapse of a whole society ... . We are dealing here with a famine of biblical scale—a great destruction.
>
> (Communications Directorate Fisheries and Oceans, 1993, see also Rogers, 1995)

The cod fishery in the North Atlantic still has not recovered even after it was shut down in the 1990s. This is a story of exploiting the marine system to promote social and economic welfare, but not knowing exactly where the boundary was to protect the systems that supported that same welfare. Worse, the cod collapse is a grim model for failing fisheries around the world. Since World War II, we have doubled the effort to land fish, but because of over-fishing, this effort is landing half the fish the same amount of effort would have achieved in 1950. Many people rely on fish protein, and to feed this demand, more and more boats are chasing fewer and fewer fish, going deeper and deeper into the ocean as the more productive and rich coastal fisheries are successively exhausted. While coastal fisheries have been serially depleted, reducing fishing effort before it is too late can stabilize or reverse losses of revenue and fish populations that have occurred from serial depletion, demonstrated in isolated cases in Norway, Iceland, the US, Canada, Australia, and New Zealand. For these and related synergistic problems emerging in global environmental change, Norgaard and Baer write, "Our future as a species depends on our ability to grapple with the complexities that arise in interactions between social and environmental systems" (Norgaard and Baer, 2005).

The story of the cod represents several recurring issues in sustainability we will explore in this chapter. Clearly, the social and ecological systems of the cod have been deeply connected and sustainability itself is a systems problem, discussed later. Second, the welfare of the fishing communities increased with more cod, and this allowed the community to have more boats to prosper even further, but this growth and prosperity fueled by fish undermined the very source of welfare for Newfoundland and, when the fish were lost, the prosperity and welfare of Newfoundland was also undermined. Further, even though a definition of sustainability between fishers and scientists was contested before the collapse, there is evidence of durable and lasting relationships—First Principles of Sustainability—worth exploring.

THE PROBLEM STRUCTURE OF SUSTAINABILITY

While details differ in specific problems, such as the difference between maintaining healthy fisheries and soil fertility, sustainability has a consistent architecture. This architecture is what we will call the **problem structure of sustainability**: all organisms have a metabolism that requires consumption and disposal of energy and matter provided by ecological systems; but, consumption disturbs the very ecosystems necessary for a healthy metabolism to begin with and therefore the life of the organisms over time.

Related to human sustainability, all individuals, communities, civilizations, and the world in total need to consume matter and energy and dispose of waste in order to survive. This is the metabolism of societies, and this metabolism is dependent on their surroundings, the land, water, and other living things to maintain their metabolism. If this metabolism disturbs their surroundings enough, either through consumption or polluting, then the very source of the metabolism is interrupted and the organism or organization cannot continue without finding new sources of matter and energy. Societies around the world need to grow food to eat well, but if the soil is used too much and there is no fallow or rest for the soil to regain nutrients and fertility or it is polluted, this welfare is undermined. Ironically, in a simplistic way, eating very well from planting too much may exhaust the soil and lead to hunger.

Now, global humanity is connected through markets, which provide alternatives to local soil exhaustion, but may put pressure on distant soils. This is a very important complication to sustainability. Since we have global agriculture and industry, local consumption and pollution may be caused *or* relieved throughout the larger network of production, trade, and consumption. This large network also is difficult to see through, which means that consumers often cannot know what the impacts are of their consumption and waste.

The consistent problem of sustainability is the contradiction between needing to consume and the damage that consumption can bring in a globally connected set of social, economic, and ecological systems.

Note, however, that the problem structure of sustainability is not the same as the problem of scarcity, which was central to the sustainability debate in the 1970s. Scarcity is not a good measure for sustainability guidelines because we fail to be interested until the system or resource in question is totally consumed, or nearly so. Scarcity also does not concern itself with how cycles and systems produce the resources, like water, that become scarce. Focusing on scarcity is like mopping up water on the floor without stopping the leak that produces the problem because it is blind to the larger causes. In other words, a concern for scarcity simply does not take account of the complexity within which modern societies operate. Still, there is a lot of uncertainty about the concrete boundaries in these systems, and this leaves a lot of contested terrain in the politics of sustainability.

CONTESTED TERRAIN

There are no uncontested positive definitions of sustainability. A positive definition is one that declares what sustainability is, whereas a negative definition would say what sustainability is not. In some cases it is easier to know when something is not sustainable than when it is. Part of this problem comes from the different types of uncertainty. If I *know* I have $100 in the bank and I remove $20 every week without replenishing it, this process is only possible for five weeks. If the goal is to have any money at week six, this process is patently unsustainable. I know how much money is in the bank, how much is deposited, and how much is removed. I do not have to guess how much other people will take, or if the bank will be randomly robbed because the money is insured. Global sustainability that makes the world work for 100 percent of humanity is much harder to define, and there are a lot more uncertainties about resources and system limits, what others will do, and what random events may occur along the way.

To define sustainability positively is also difficult because sustainability is an **essentially contested** concept. Michael Jacobs (1999) explains that essentially contested concepts have two levels of meaning—one aspect is agreed upon but vague. The second aspect is not agreed upon. This second aspect concerns interpretations of the term for practice. Like democracy, liberty, and justice, we know the basic ideas, but the real political conflicts come at the second level when the ambiguous idea is translated into real-world decisions. There is agreement about the basic, but vague, meaning of sustainability. It is agreed that the long-term continuity of societies depends on preserving critical ecological life support systems, and Jacobs writes, "Its first level meaning is now given ... the core ideas are fixed and cannot now be changed through rational argument."

However, in practice we have to decide how to balance our material metabolic needs to live (economics), how to live within ecosystems without tearing them asunder (ecosystems), and how to live with each other (equity and justice), sometimes referred to as the **three "Es."** This means we need to know "how much" of each measure we need to continue and improve our prospects as individuals, nations, and a species. The three "Es" here are thought to be like legs on a stool, which means that one E cannot be deemed more important than any other. Jacobs argues that each of the three "Es" makes up critical "fault lines" of political and ideological debate that are persistent.

First, while we may all agree that environmental protection is needed, there is disagreement on how much is needed and the appropriate way to go about it. Second, most people around the world, when polled, agree that equity is important to a sustainable world (Leiserowitz, Kates, and Parris, 2006), a sound economy, and a just society, but there are trenchant ideological differences as to what rules to put in place, how much should be redistributed to the less well off, and which less well-off communities are relevant (non-humans, local, national, or global poor).

In addition, those who focus more on economics tend to adopt **weak sustainability** approaches that require less transformation of society because these advocates presume problems will be solved by the market and future technology; but, those who focus more on ecology, adopt a **strong sustainability** approach that requires a more dramatic transformation of society's values and practices. Weak sustainability argues that we can continue to consume ecological goods and services at a growing rate, whereas strong sustainability insists on strict limits to consuming ecosystems.

These fault lines privilege specific approaches to sustainability. In economics, maintaining revenue sources is the focus, and in ecology the goal is to maintain the functions of ecosystems.

However, even though we cannot report a positive definition for sustainability, there is strong support in the literature that identifies sustainability as an evolutionary process guided by *principles*:

> Interpreted this way, sustainable development is a dynamic concept. ... [and] the best that is likely is to be possible is to articulate *general principles* to assess the relative sustainability of the society or the economic activity compared to earlier states or economic activities.
>
> (Folke and Kåberger, 1991, emphasis added)

One widely cited set of principles comes from Herman Daly (Goodland, 1995), an economist, found in Box 2.1 below.

BOX 2.1: DALY'S OPERATIONAL PRINCIPLES OF SUSTAINABLE DEVELOPMENT

a Renewable resources should not be harvested beyond their regenerative capacity.
b Non-renewable resources should not be consumed faster than substitutions can be produced (though any consumption of non-renewable resources is, by definition, not sustainable).
c Sinks should not be used beyond their natural assimilative capacity.

While we do not have an indisputable definition of sustainability, there is a remarkable consensus on some *principles*, or standards, for sustainability.

FIRST PRINCIPLES OF SUSTAINABILITY

Although ways of life differ around the world, every society requires essential support systems of ecosystem services for social and material wellbeing. As a universal requirement, this provides us with the basis for general principles of sustainability. Indeed, Davison (2008, emphasis added) argues that we might think of sustainability as something like an atom:

> Imagine *principles* of sustainability as a tightly aggregated nucleus around which orbit only loosely aggregated goals of sustainable development. The small area of agreement established by the ideal of sustainability is nonetheless sufficient to bind together a wide constellation of diverse sustainable development objectives.

From the Millennium Ecosystem Assessment (MEA), we know that humanity has benefited from global environmental resources and ecosystem services, but we have also initiated unprecedented global environmental changes more quickly than ever before—human activity has degraded about two-thirds of the entire world's ecosystem services.

MEA reports that the "bottom line" is that,

> At the heart of this assessment is a stark warning. Human activity is putting such strain on the natural functions of Earth that the ability of the planet's ecosystems to sustain future generations can no longer be taken for granted. The provision of food, fresh water, energy, and materials to a growing population has come at considerable cost to the complex systems of plants, animals, and biological processes that make the planet habitable.
> (Millennium Ecosystem Assessment, 2005b)

The MEA makes it clear that we are "living beyond our means" at the planetary level. The MEA is but one corroboration that eco-system services are required for society, and if these services are degraded enough (how much is not easily known) the world's societies will suffer crises and perhaps collapse. How society deals with such a crisis varies, but when, say, climate changes alter the availability of water, the people without enough water to grow food and live meaningfully will face a crisis of subsistence. Indeed, research shows that deteriorating ecosystem services now pose a series of health threats, especially to the poor. Myers and Patz's research in this area brings them to conclude that, "These threats include increasing exposure to infectious disease, water scarcity, food scarcity, natural disasters, and population displacement. *Taken together, they may represent the greatest public health challenge humanity has faced*" (Myers and Patz, 2009, emphasis added).

Consequently, the maintenance of ecological systems and their ability to provide basic living conditions for 100 percent of humanity are the core P1 of the **First Principles of Sustainability** in that they provide necessary *but insufficient* conditions for the continuity of societies.

In addition to the first ecological principle noted above, a sustainable society must avoid the more difficult "**Normative Failure**." The word "normative" means "what should be," and enacting social rules that reflect what "should be" involves the more contested social issues of ideology, morality, values, ethics, governments, and institutions, among other disputed conditions. Still, some governing systems will fail to institute the normative injunctions against P1 requirements, and therefore they fail to provide the normative constraints to ensure basic needs. These principles are strongly represented in the sustainability literature. Thomas Princen, for example, elaborates Normative Failure as a *principle* of sustainability:

> Under ... conditions, namely [of], environmental criticality, a different set of *principles* is needed, a set that embodies social restraint as the logical analog to ecological constraint, a set that guides human activities when those activities pose grave risks to human survival.
>
> (Princen, 2003, emphasis added)

Princen defines environmental criticality as:

> environmental threats, problems characterized by irreversibility and non-substitutability, threshold and synergistic effects ("surprise"), long time lags between cause and effect and, consequently, limited predictability and

manageability. Climate change, biodiversity loss, topsoil erosion, persistent toxics, and declining freshwater availability are examples of such threats.

In the context of sustainability, we are charged with at least two normative responsibilities: guiding how we treat biophysical life supports, and guiding how we treat each other—both of which are necessary for the continuity of any society. A society that allows or gives itself permission to overwhelm ecological spaces, fails to adapt to vulnerability, or that produces effectively virulent social diseases suffers from "Normative Failure." A sustainable society must be able to adapt to ecological changes, and avoid fatal Normative Failures. Unfortunately, when we learn about collapse, we will see that failures at the civilization level always occur in a complex series that stop us from making simple prescriptions that would assure avoiding such failure today.

BOX 2.2: THE FIRST PRINCIPLES OF SUSTAINABILITY

P1

Without ecological life supports, there is no society. This relationship is immutable. A sustainable society must maintain the integrity of Earth systems and cycles that provide critical life supports. The threshold between a system with integrity and resilience is non-linear and the point of change often is unpredictable. Ecological and social systems are profoundly interdependent and changes anywhere in any system cause other, often unintended, consequences elsewhere.

P2

What kind of society that grows in an ecological space is a value-based question, but sustainable societies must observe normative constraints:

a The social system will not be sustainable if it undermines ecological life supports (principle of accountability and restraint).

b The social system will not be sustainable if it sufficiently militates against itself or is annihilated by others (principle of justice).

c The social system must be adaptive to challenges and changes to avoid evolving vulnerabilities (principle of foresight).

In summary, sustainability is achieved through the enduring maintenance of what I will hereon refer to as **First Principles** (P1 and P2). Violating P2(A–C) constitutes a Normative Failure. Princen notes that among the normative principles of sustainability, three are most important. First, societies must observe restraint because ecological systems are constrained; second, societies must avoid exporting risks to others; and, third, societies must not escape responsibility for the creation of environmental problems because

> technological innovation and market manipulation skew the benefits and costs of economic activity, to create the illusions of environmental progress (e.g., local pockets of pristine and healthy environments, especially among those who can buy their way out of degraded environments) while vast areas around the world are degraded and huge waste sinks such as the oceans and atmosphere are filled.
>
> (Princen, 2003)

Note how Princen's normative principles indicate that empire or exporting risks is not consistent with sustainability—while a more powerful country might export its risks and hazards (like pollution) to make themselves better off, overall this is not sustainable for a number of reasons. These criteria are "normative" because P2 requires that a society or network of societies value and do the right things. P2 is the source of most debates about sustainability, because values and behavior are permanently in contest and are inherently political.

However, P1 is the small area of agreement about sustainability. P1 is where the notion of sustainability loses its relativism. For example, Fischer et al. (2007) write:

> Sustainability is not a relativistic concept because the biophysical limits to sustaining life on Earth are absolute. Societies cannot exist without a

functioning life-support system, and economies can only flourish within a functioning social system with effective institutions and governance structures.

Indeed, Williams and Millington (2004) indicate that, while there are important contradictions in how sustainable development is discussed,

the starting point of much of the sustainable development literature, albeit more often implicit than explicit, is what we here call the "environmental paradox". For nearly all commentators on sustainable development, this means there is a mismatch between what is demanded of the Earth and what the Earth is capable of supplying.

Fischer et al. (2007) note that, given key changes like climate change and global biodiversity loss, humanity is driving farther and farther off the interstate and away from sustainability:

For the first time in human history, our activities are so pervasively modifying our own life-support system that the ability of the Earth to provide conditions suitable for our species to thrive can no longer be taken for granted.

SUSTAINING SYSTEMS

The above makes clear that sustainability is a problem about maintaining various *complex* systems: social, economic, and ecological systems in particular. This section will explain how complex systems tend to work.

A **system** is an organized set of parts that create a larger unified whole that neither one of the parts could have produced alone. Thus, the essence of any system is in the way the parts affect each other. A **complex system** is one that has many internal parts and many relationships between these parts, so that changing one part produces mostly unpredictable results. Social systems and ecosystems are complex.

The **Adaptive Cycle** is a model of how complex systems change over time, and it links ecosystems and social systems in "never-ending adaptive cycles of growth, accumulation, restructuring, and renewal" (Holling, 2001). Successful adaptive management experiments with natural resources in ways that encourage learning, new opportunities (novelty), choices, and fosters resilience; however, exploiting social-ecological systems without any consideration for the future leads to narrowed choices, inflexible options, and collapse (Folke, 2006).

This book will treat sustainability from a systems perspective given these considerations. In fact, sustaining planetary civilization requires a system of systems mindset.

Norton (quoted in Turner, 2005) offers five relevant axioms consistent with the Adaptive Cycle and sustainability:

1 "'The Axiom of Dynamism'—nature is a set of processes in a continual state of flux, but larger systems change more slowly than smaller systems.
2 'The Axiom of Relatedness'—all processes are interrelated.
3 'The Axiom of Hierarchy'—systems exist within systems.
4 'The Axiom of Creativity'—processes are the basis for all biologically based productivity.
5 'The Axiom of Differential Fragility'—ecological systems vary in their capacity to withstand stress and shock."

These propositions apply to social systems as well. Sustainability requires that integrated human and non-human systems operate both now and in the future (Axioms 4 and 5). Naturally, if each generation maintains these systems, then the long-term sustainability of the human species is set for a very long time. However, because these systems are dynamic, external disturbances other than humans also affect them, and therefore all of these systems will change over time (Axioms 1 and 2). Sustainable management of the Adaptive Cycle means building societies that are able to live through these changes, and perhaps even flourish in the novelty that these changes bring. If our societies do not have enough flexibility, savings, or memory to live through inevitable change, then we face oblivion, just as Buckminster Fuller warned (Axiom 5).

Social-ecological systems are linked across **scales** (Axioms 1 and 3). Scales are linked systems across dimensions of time and space. Scales are nested together with local and small time frames, e.g., hours or days, to the planetary and large time scales, e.g., geologic eras, with each smaller scale feeding into next higher scale.

An example of a local small scale is the hourly weather in a town or city, while at the other end of the scale, we have planetary climate that is slow to change—until a breaking point is crossed. The precise moment of this "breakpoint" is called a **threshold** of change and it is a place where slow mounting changes build up and "deep uncertainty explodes," creating massive change (Holling, 2003). Typically, we cannot predict precisely where or when the threshold exists. Systems can persist and even be improved by these events, and persisting under these changes is critical for worldwide sustainability. World civilization and its subsidiary societies will certainly move through thresholds of change, as it has in the past. The question is what will that threshold bring and will these societies be able to live through this change and re-organize?

All change is not bad, which is good news, because changes occur with or without us. Breakthrough events can be sustainable if they bring renewal, like when Gandhi's social movements triggered the eventual de-colonization of India, or these changes can be the final concluding death-throes of a society, or all societies (e.g., under nuclear Armageddon).

Because each system is linked to other systems (Axioms 2 and 3), what happens in one locality affects other localities through the matrix of connections. One community may live within the limits of its systems, others may not, and this dynamic makes sustainability a permanent transnational problem.

PANARCHY AND THE ADAPTIVE CYCLE

One groundbreaking contribution to sustainability comes from the Resilience Alliance, and one of the most important publications of their research is the 2002 edited volume, *Panarchy: Understanding Transformations in Human and Natural Systems*. The Resilience Alliance is a program of the Beijer International Institute for Ecological Economics, Swedish Royal Academy of Sciences in Stockholm. The Adaptive Cycle described here comes from numerous empirical studies around the world (Scheffer et al., 2001). At the root of their contribution is the idea that sustainable systems are resilient systems—they can recover from and withstand inevitable disturbance.

The notion of the Adaptive Cycle originates with ecologist C. S. "Buzz" Holling, who dislodged the mainstream ecological expectation that ecology has a single equilibrium where it is stable in a single state. Scheffer et al. (2001) write:

> Nature is usually assumed to respond to gradual change in a smooth way. However, studies on lakes, coral reefs, oceans, forests and arid lands have shown that smooth change can be interrupted by sudden drastic switches to a contrasting state.

Holling showed that ecologies operate through multiple stable states, not one. Some of Holling's initial research in this area was on the spruce budworm. This work was essential to the timber industry because it affected the practicality of how many trees to take. There was a relatively easy way to figure out the growth patterns of spruce and therefore identify the "surplus" of trees from this growth. This allowed the industry to think they knew the maximum amount of trees they could take without undermining the forest's productivity, referred to as the **Maximum Sustainable Yield** (MSY). However, the managers did not know how to work with outbreaks of budworm, which is a moth in the larvae stage that eats the new needles in pine forests. The budworm works in cycles where the population of budworm is controlled by birds, when the birds can easily find them. However, when the forest grows enough, the budworms are better hidden in the foliage and the birds cannot hunt them as efficiently. The budworm then goes through an explosive population growth. In order to avoid the explosive growth that would eliminate a large part (~80 percent) of the forest, Canadian timber managers had been attempting to control the budworm with pesticides, but then found themselves locked into this approach because, as the forest matured, if they released the budworm from the pesticide it would devastate the forest. Clearly the combination of the budworm cycle and the harvest of trees had to take each other into account.

Holling demonstrated the cycle of the budworm to the managers so that they could create a "patchy" pattern of forest as well as use better timing for the pesticide to avoid a collapse of the forest. In this way, he developed a more sustainable approach to timber cutting and pesticide use.

To understand the central propositions of the Resilience Alliance, we need to understand a few key concepts and terminology that influence the context for sustainability throughout the rest of the book.

The first concept is that of **social–ecological systems**. Social-ecological systems refer to the coupled/linked complex systems of societies and ecologies. Take notice that we are thinking specifically in terms of systems. Systems theory assumes that a system is constituted by the relationships between parts. The relationship of each part to other parts creates a larger whole. Some systems are more complex than others. We may have a car made up of parts that affect the overall ability of that car to carry people, but these parts are limited and rather simple compared to "society," for example. Social and ecological systems are constituted by a very large number of relationships. There are so many relationships between people, non-humans, and the non-living environment, the impact of changing one part of a society or an ecosystem is very difficult to anticipate because there are chains of reactions that occur within the system. One change in society or ecology may initiate an unintended cascade of changes within the system.

If the cascade fundamentally changes the order of the system, the system has experienced a catastrophic shift (see Scheffer et al., 2001). Another way of saying "order" is "regime" or "state," and so catastrophic shifts are also **regime shifts** or state changes of the system from one stable state to a different state.

Resilience is the capacity of a system to experience a disturbance, and then return to its original state, avoiding a regime shift. There is a growing sense that the true goal of any program for sustainability should really be to build resilience, though this goal is contested because having a resilient system is only good if the system is "good." There may be social systems that many people would or have been very glad to see pass their thresholds, such as the Third Reich, which reminds us that in some cases sustainability is not preferable. That said, much of the work toward global sustainability has focused on, for example, maintaining the integrity of vital ecological systems where just social systems can be possible.

The opposite of resilience is **vulnerability**. Vulnerability is organized by the original condition of the system, exposure to disturbance, and the sensitivity of the system to these disturbances (Luers, 2005). The starting point of the system is important because it is the position of the system relative to a potential state change. We can think of the starting point as something like the distance between us and the edge of a cliff. If we are on the edge, we are more vulnerable than if we were far away. Let's use the example of the orange roughy (*Hoplostethus atlanticus*), a fish that can live 150 years, but takes 30 years or more to become mature enough to reproduce. This fish is caught in the Southern Ocean region. Given the fish's exceptionally long sexual maturity, the population is more susceptible to depletion than other fish species that are able to repopulate more quickly. This means that a disturbance can take the fish closer to collapse than other fish populations, all things being equal. In addition, the more that orange roughy are exposed to fishing (exposure to the disturbance), the more vulnerable they are to collapse. In the 1980s, the fish came into popularity as a tasty meal, internationally. Fishing pressure increased, and the sensitive species was exposed to more disturbance that built up quickly, causing a severe contraction of its numbers, collapsing the species to about 20 percent of its original numbers of the 1970s (Clark, 1999).

Thus, the ultimate goal of learning from the Adaptive Cycle is to build resilient social-ecological systems that avoid unwanted regime shifts. The cod of Georges Bank experienced a very unwanted regime shift when it collapsed, as did Rome when its empire collapsed. The Adaptive Cycle provides us with a conceptual map of how social-ecological systems change based on internal dynamics and external pressures through four phases identified by Holling (2003):

The Fore Loop of the Adaptive Cycle:

1. Growth (r) by exploiting resources that builds structure and high resilience (see lower left section of Figure 2.1).
2. Conservation (K) as growth slows and the system becomes more brittle and vulnerable to outside stress.

The Back Loop of the Adaptive Cycle:

3. Release (Ω) where a disturbance brings the collapse of system and its defining structure.
4. Reorganization (α) can occur after collapse if there is enough energy and matter, which may allow for a new phase of growth, potentially in a new system.

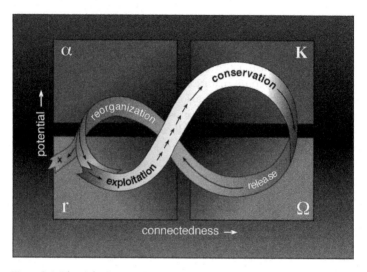

Figure 2.1 The Adaptive Cycle

Ecological, social, institutional, and social-ecological systems appear to follow the Adaptive Cycle (Walker et al., 2006; Walker et al., 2002). **Panarchy** is the term describing the geographical and chronological hierarchy of Adaptive Cycles that are nested within each other. Localities are nested within regions and regions fit within a global Earth system, etc. Processes between geographic spaces are linked across scales, such as the way that weather (short time scales) is situated within and affected by longer time scales of climate. So, small systems like genes or organs exist within individuals, who exist within families, who exist in communities, and regions, and continents, all of which exist in the planetary system. All of these systems are regulated by positive and negative feedbacks. Positive feedbacks accelerate change, negative decelerate change, and all systems can fail. Sustainability, then, is built on the integrity or strength of the systems found at each scale. Higher scale failures will deeply impact the lower scales that exist within the higher orders. We might have families that go bankrupt, as many did during the Great Recession starting in 2008, and avoiding this problem is a problem of sustaining the family system. This is different from

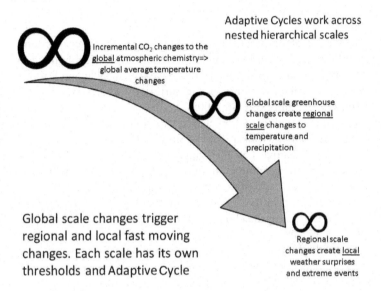

Adaptive Cycles work across nested hierarchical scales

Incremental CO_2 changes to the global atmospheric chemistry=> global average temperature changes

Global scale greenhouse changes create regional scale changes to temperature and precipitation

Global scale changes trigger regional and local fast moving changes. Each scale has its own thresholds and Adaptive Cycle

Regional scale changes create local weather surprises and extreme events

Figure 2.2 Panarchy and nested Adaptive Cycles linked across scales

sustaining the world capitalist system, which is the dominant market system connected around the world; and, sustaining the world capitalist system is different than sustaining the world population. But each of these systems—from family bankruptcy to the human population—are connected in the larger project and discussion about global sustainability.

Therefore, Panarchy looks something like Figure 2.2, where larger scales have larger spaces in which to change, and therefore take longer to change, at least in the Fore Loop.

Within larger scales, like the global climate system, there is an enormous amount of space and energy, and changing climate in the Fore Loop is a slow time scale. However, as this larger scale changes, it affects the smaller scales that fit within the global scale—and these changes occur faster and are over smaller space. And, at the smallest local scales we have smallest space and fastest time changes as in daily changes in weather. Since weather is nested in the larger climate system, it is affected by that climate. If climate changes, as it has been for the last 100 years or so, then weather will also change.

When the global system is changing, it may drive local systems toward surprising changes, and climate change will probably result in very serious local surprises. If adaptive governance, including learning from our mistakes, does not occur, surprises become crises (Folke, Hahn, Olsson, and Norberg, 2005).

In addition, within each cycle, we are concerned with slow and fast changes. Often, we can control slow variables but not fast ones. Within the Adaptive Cycle, it appears that catastrophic shifts are triggered by an accumulation of changes from slowly increasing disturbances. Slow additions of CO_2 accumulate in the atmosphere, but at some point in this accumulation, there is enough CO_2 in the atmosphere to warm the climate system to a point where catastrophic shifts may occur.

If you think of a stable system as something like a person in a canoe, as Scheffer explains, and you want to see something in the water—there is a critical transition at some point where you lean over too far (or a wave comes, or you sneeze, etc.) and the canoe tips over (Scheffer et al., 2001). The system passes a threshold and goes through a "catastrophic" shift that, in ecology, will never go back to its exact original state. Similar systems may emerge, but the exact system will not re-emerge.

Ecosystems go through abrupt and catastrophic shifts, determined by slow changes that mount. However, because there are so many relationships to a vast number of parts in social-ecological systems, the point at which society, ecology, or social-ecological systems will "tip their canoe" is, for all practical purposes, impossible to know with precision. We can theorize about the way in which these systems exist in Panarchy and their distinct, interconnected phases, but the systems are too complex to predict at what point we cross the threshold, at which we go from conservation to release—or slowing growth to collapse. However, if slow variables control the accumulation, then humans can behave in a way that reduces this accumulation.

We can control CO_2 emissions to some extent, but we cannot control the consequences of climate change, including the point at which thresholds of change exist. As the climate warms generally at the planetary scale, it will cause non-linear changes across smaller, embedded scales that may or may not even resemble warming because these global changes will cause regional variations, and surprises from changing the relationship of the parts. For example, as climate has warmed, there has been a resulting increase in bark beetle infestations across the inter-mountain West of the United States, perhaps because the warming interrupted the pulses of populations of bark beetle because colder weather normally interrupts their population's growth phase. Less precipitation has also made the trees more vulnerable. The combination has resulted in a breathtaking loss of trees and whole forests. Thus, we could control CO_2 emissions, but at the time of this writing we cannot control bark beetle outbreaks unless you bag and chip every infested tree across millions of square miles.

Other important thresholds appear to be very important in climate change. There is a vast storage of methane at the bottom of the ocean in clathrate belts, and there is a huge amount of methane and CO_2 in the tundra. As climate warms, one ominous threshold appears to be the point at which oceans and tundra warm enough to release these greenhouse gases. The release of these stores will very likely create a regime change, and create a whole new climate order and cascades of changes across smaller scales (Overpeck and Cole, 2006). Notice that the bark beetle issue is located at the regional level, while clathrate methane would be a global issue— thus, because scales are linked across scales, we have many canoes! As there are several scales, there are multiple thresholds for the various nested systems.

Again, thresholds of change are governed by slow additive variables, but trigger non-linear (fast) sequences that result in a state change of the system itself. The implications for sustainability are clear. Experiencing a state change in the social-ecological systems, say agricultural conditions or soil fertility or renewable resources, may have grave implications for people who depend on these ecosystem services. It is possible that the Back Loop of the Adaptive Cycle explains what it looks like when the larger social-ecological system violates core principles of sustainability (Holling, 2003).

Thankfully, research consistently has indicated that there may be early warning signs of an impending regime shift. These early warning signs, however, may be difficult to monitor in the real world because they require a high degree of observations in large systems like forests, rangelands, and freshwater and marine systems. These observations involve watching for oscillations in basic functions and productivity, such as the color and production of chlorophyll, that can be evident far in advance of unwanted shifts in system states. While onerous, this kind of observation—and then follow through—in human responses may be an expensive but necessary activity in the Anthropocene.

The Adaptive Cycle is not a universal metaphor, and there are systems that go through some of these cycles but not all of them, yet the metaphor is applicable to many systems and consequently it is sometimes referred to as a "meta-metaphor." The Adaptive Cycle provides both a pragmatic and compelling theory for several issues related to sustainability.

For example, this heuristic theory demonstrates why MSY policy tends to oversimplify system dynamics even if MSY is implemented with honesty—that is, the policy makers really set harvest rates of a renewable resource at levels that can be sustainably replenished, all things being equal. This is because MSY usually assumes that the human harvest rates are the only thing affecting the resource, and it ignores the system in which the resource exists. In the case of the spruce budworm, if birds are ignored in the process of harvesting, managers may set a perfectly reasonable MSY based on known rates of regeneration of trees. However, if the harvest and removal of trees crosses the threshold where there are not enough trees to support the birds, then they will not eat and suppress the budworm. The budworm then would be released from predation, and, without other devices like pesticides, could devastate the forest. Thus, MSY is often not sustainable or resilient! Central to our goals in building sustainability, we learn from the work of the Resilience Alliance that sustainability and attempts for P2 will maintain the integrity of whole ecological and social systems within which specific resources exist. In this way, simply trying to avoid scarcity is too myopic of a measure because it only thinks about the amount of fish left in the ocean, without realizing that the fish live in complex food webs and a complex water column in the ocean. If the marine systems are ignored, then the support systems for the fish, and therefore people, may be weakened to a point where both decline or even collapse.

Sadly, however, natural resource management schemes have been slow, if not deaf, to this change in science. For example, MSY has been challenged in mainstream fishery science at least since the 1970s, but mainstream policy tends to still use MSY domestically and internationally (see for example Larkin, 1977).

In addition, the Adaptive Cycle gives us important insight into the nature of collapse. The Back Loop (see p. 51) tells us there are slow variables that lead up to sudden changes, and managing for sustainability means managing for resilience, or against vulnerability. And, just like in *Limits to Growth* (LTG) (Meadows, Meadows, Randers, and Behrens, 1972), the point before collapse is stable—the opposite of collapse—and until collapse happens, critics may have evidence that everything is going well.

Less vulnerable systems will learn and act based on lessons of the past. More resilient systems will have and use long-term memory, and will take advantage of crises as moments to bring novel approaches into practice (Folke et al., 2005). We learn from this approach that as early resources are exploited in a system, its growth is easily predicted, like the growth of worldwide fish catch as the world industrialized its fleet in the 1960s and 1970s. But, as growth slows, just as it has in worldwide fish catch, the accumulation of structure and dependency, such as revenue needed to pay for new boats, means the system becomes less flexible. Thus, even if we know fish populations are down, high fishing effort may continue because fishers need to pay their bills. This is the point where thresholds exist, and indeed, we have seen a systematic rise in the rate of over-exploitation of fish that matches the growth in fishing effort, and the now legendary collapses of key species like cod.

Since there are early warning signs that exist in these systems, and some regime shifts will clearly be unwanted by society, monitoring for these regime shifts may play a role in building resilient social-ecological systems. So-called "**adaptive governance**" would make societies less vulnerable and more resilient to the dangerous prospects of rapid changes at the local level that are expected when slow changes are made at the global level (Luers, 2005). Most of the time, adaptive governance will require human activities to reduce slow but mounting disturbances to the system, like harvesting trees or fish, creating erosion, or adding greenhouse gases to the atmosphere while preparing for regional and local threats that are already created.

Thus far, there are few examples of this kind of adaptive governance, but academics have recognized the need for adaptive governance as global environmental changes have become more and more pressing and uncertainties multiply across regions, issues, actors, and needs (Folke et al., 2005).

ENDPOINTS

WHAT DO WE KNOW?

Thinking about sustainability has matured significantly since the 1970s, when the central concern was running out of minerals, like oil, or water, to the point now where those concerns still exist, but more importantly we are concerned with maintaining critical life support systems and cycles that produce oil and water, etc. Further, we now understand that these cycles and systems can and have withstood enormous pressures from human activities, but that these systems reach breaking points where very fast, irreversible, and profound changes occur that alter the way that whole system works.

We know that sustainability is a structural concern, and that environmental policies and activities that may be "green" or environmentally beneficial do not address sustainability unless they address the systemic nature of maintaining social-ecological systems—that they work in favor of First Principles.

These First Principles are well represented in the scholarship about sustainability, but the practice of sustainability is a normative and contested question that forces us to ask "what is a good world?" and "what is the best way to live that favors current and future generations?"

CRITICAL CONSIDERATIONS

If we envisioned the world as a set of ecosystems along the Adaptive Cycle, where on that cycle do you think we are, for the most part? How do you justify this assessment?

What kinds of problems do you see in the formation of sustainability First Principles?

What kind of world would you like to see? What kind of governments is it made up of, ideally?

What kinds of expectations do you think would exist in your "good world?"

WHAT DO YOU THINK OF THESE SUSTAINABILITY SOLUTIONS?

1   Governments around the world fund and deploy early warning systems to systemic change in critical life support systems. When signals of changes are detected, emergency law-making meetings are called to make decisions, based on the level of government that is relevant—if there are international systems problems, then emergency meetings of the UN Security Council would meet to protect the security of all people; if it is a city-level problem, city officials would meet. This would require an extensive network of monitors to detect specific changes.

2   A labeling system is adopted for all goods that indicates any effects on ecosystem goods and services the production of that good may have. The label includes a map of ecosystem services that the good is connected to, and any problematic connections, say to water depletion or climate change, would raise the price of that good in a way that would truly represent the overall cost, ecological and economic, to society.

WHAT DO YOU THINK OF THE FOLLOWING SYLLOGISM?

Premise A: The only way to know where a threshold point in a complex system, like a fishery, is is to overshoot it.

Premise B: Understanding where these breaking points in systems like fisheries are critical for sustainability.

Conclusion: Policy makers, such as those managing fisheries, must exceed system limits, e.g., permit too much fishing, on purpose to know where the thresholds are.

FURTHER READING

Gunderson, Lance H. and C. S. Holling (eds.). (2002). *Panarchy: Understanding Transformations in Human and Natural Systems*. Washington, DC: Island Press.

This book provides a strong overview of nested complex adaptive systems (Panarchy) in social-ecological terms. Chapters from leading systems thinkers and founders of the Resilience Alliance provide important ways to think about sustainability problems.

Norton, Brian. (2005). *Sustainability: A Philosophy of Adaptive Ecosystem Management*. Chicago, IL: University of Chicago Press. Norton, a philosopher and long-time thinker about sustainability, provides enormous insight into dynamic systems, in terms of knowledge systems for survival, responsibility, and other key insights.

SUSTAINABILITY PRINCIPLES

Bosselmann, Klaus. (2008). *The Principle of Sustainability: Transforming Law and Governance*. Aldershot, UK: Ashgate Publishing.

Dresner, Simon. (2008). *The Principles of Sustainability* (second edn.). London: Earthscan.

Princen, Thomas. (2005). *The Logic of Sufficiency*. Cambridge, MA: MIT Press.

Wilkinson, Roger and John Cary. (2002). "Sustainability as an evolutionary process." *International Journal of Sustainable Development* 5(4): 381–91.

These sources are just a few examples of the literature discussing principles of sustainability.

Bettencourt, Luís M. A. and Jasleen Kaur. (2011). "Evolution and structure of sustainability science." *Proceedings of the National Academy of Sciences* 108(49): 19540–45. doi: 10.1073/pnas.1102712108. This research article provides background to the growing consensus positions in "sustainability science," which is the rigorous investigation into causes of sustainability problems and their potential solutions.

ENDURANCE AND RUIN
AN ECONOMIC MEMOIRE

MAP OF THE CHAPTER

This chapter discusses economic thinkers who deal mainly with the problem of scarcity as it relates to sustainability, but the background storyline is how an overly pessimistic or optimistic view of scarcity and abundance foils reasonable problem solving. An overly pessimistic approach may not generate enough hope that efforts to solve a problem, like policies or foresight of planning or restraint, will be worth it; meanwhile, an overly optimistic approach may generate so much confidence that problems will solve themselves through magic that there is no need to put effort into solving the problem. Both extremes may confuse or paralyze social action necessary to solve complex interdependent problems of sustainability. The pessimistic approach is proposed by one of the more important thinkers in sustainability, Thomas Robert Malthus. Malthus proposed that the checks on population will always precipitate cyclical collapses, especially of poor populations. The optimistic approach is the response Malthus received from Godwin who proposed that humanity would eventually triumph over the material world and that human destiny would overcome any limits found in the world. Then there is Ester Boserup who argued that population pressured people to adapt their food systems, and grow more—the reverse of Malthus' trap. Finally, we will discuss the related idea of sustainable development as it was proposed by the Brundtland Report in 1987.

We begin this chapter with two portraits, one of the Auroville Earth Institute in Chennai, India, and one of development of the tar sands in Alberta, Canada.

Satprem Maïni is the Director of the Auroville Earth Institute, whose mission is to research an earth-based architecture. The general idea is that buildings can be made sustainably with local earthen material by compressing soil and designing these compressed blocks so that they can lock together and provide inexpensive, energy efficient living space while training local workers to become skilled masons in the process. Since 1989, he has trained over 8000 people in 75 countries in this way.

The process begins when soil is excavated from the same local space for the building, and then it is tested and stabilized to make it stronger. It is then compressed into blocks with presses. These blocks are the material for the building, which can be several stories tall. Then, the excavated space is restored ecologically, sometimes as a water filtration pond or other space of rich life, as opposed to a trash pit. This has the power to convert slums and other vulnerable living spaces into healthy, safe, and permanent living spaces that are a pleasure to be in because the blocks actually "breathe" and take in pollutants, keep the temperature even without a heating or cooling system, and can be built to be earthquake resistant. The institute comes to these processes with a specific consciousness of improving the lives of the worst off around the world while protecting Mother Earth:

Our Mother Earth gives us a wonderful building material, which should be used with awareness, sensitivity, and with much respect and gratitude. The Auroville Earth Institute is acting for this recognition.

(Auroville Earth Institute, 2013)

Maïni writes:

> I don't see the Earth as a formless material without consciousness, but as Spirit consciously disguised as matter.
>
> (Ibid)

This vision and commitment provides a goal of improved material lifestyles for (especially rural) poor households, improved social equity and opportunity for these households, and ecological integrity of the local environment. Projects such as these, and groups like Engineers without Borders who look to provide affordable, effective ways to improve and save lives by opening up opportunities, provide a reason to be optimistic for the future. In this future, committed and ethically conscious projects innovate new, ecologically beneficial, inexpensive, and durable opportunities to improve the lives of people who are worst off. The beauty of so many of these kinds of projects—and there are too many to mention here—is that they do not even require aid or charity, but rather the opening of opportunity where it has otherwise been suppressed. Sensible financial mechanisms and policies can support sustainably designed projects that improve the material welfare and ecological integrity of the local environment. However, often design, engineering, and development projects do not look like the projects of the Auroville Earth Institute, but more like the Alberta tar sands.

The tar/oil sand fields of Alberta, Canada, paint an entirely more pessimistic future. It is perhaps one of the "largest industrial projects in history," and represents a disturbing trend of a "growing reliance on non-conventional fuels," that create "increasing scale of environmental disruption" (Davidson and Gismondi, 2011). The shift of power to the economic sphere, where corporate interests have substantially more influence than other social and environmental concerns, is consistent with the main economic approach of the modern period since the 1970s of neoliberalism (Centeno and Cohen, 2012).

The tar sands lie under former forests, which have been torn from the Earth to mine the sand and clay soils that are soaked in low-grade oil called bitumen. In addition to strip mining to get the tar sand oil, mining companies inject steam at a high pressure that liquefies the oil and separates it from the sand. The oil then is recoverable through wells. The process has left these former forests something akin to a moonscape extending over 530 km^2 (330 square miles) of barren land with toxic retention ponds that dot the land. Hundreds of thousands of gallons of oil have seeped and spilled into the landscape, in addition to the vast habitat loss that has harmed wildlife. The reason for this project is to extract the oil for energy and produce billions of dollars in economic revenue (admittedly for a concentrated elite minority). The energy is difficult and expensive, and this indicates that oil has become valuable and scarce enough that companies and countries are willing to pursue the most expensive and difficult sources. Another example of this behavior is that rigs are drilling in deeper and much more dangerous marine waters, and it raises the specter that the world's sources of oil have or will soon reach their peak, so-called "peak oil."

Tar sands oil will emit far more carbon dioxide than even standard oil. Consequently, it fits the modern story of pursuing even the dirtiest energy sources with very high external social and environmental costs that will continue to spur climatic change. When the oil is squeezed from the sand and soil, this oil will be shipped through a controversial pipeline that is in the process of being built across the United States, which will travel over the largest aquifer in North America, the Ogallala, and other aquifers that feed Midwest agriculture. Environmental groups have protested every angle of this project, but Canadian and US tribal governments have been the most successful in raising obstacles. Tar sand development is described by some tribal leaders as a "slow industrial genocide" for regional indigenous peoples whose "ability to hunt, trap and fish has been severely curtailed and, where it is possible, people are often too fearful of toxins to drink water and eat fish from waterways polluted by the 'externalities' of tar sands production" (Huseman and Short, 2012). This development project has produced enormous revenue and an important supply of industrial energy, but scholars agree that the tar sand development is a fairly unmitigated social and environmental catastrophe akin to a "collective agreement to engage in suicidal tendencies" (Davidson and Gismondi, 2011).

Thus, on one hand we have a story of optimism that includes reductions in inequality and material deprivation while improving the local environmental functions. On the other, we have a story that engenders a more pessimistic story that the status quo political and economic values continue to be extremely powerful and are not substantially working toward sustainable energy, engineering, resource stewardship, or consumption.

In this chapter, we confront the problem of "disposition"—what attitude should we take when we think about the foreboding limits of the real world, and the genius of human imagination? In this chapter we ask just how optimistic we should be about the prospects of sustainability. On this front, some believe that the track record of countries, companies, and individuals working toward a more sustainable future does not provide much reason for hope, while some believe in an infinite human adaptability and proclivity for innovation. Part of this attitude is traced to the Enlightenment, when people like the Marquis de Condorcet and William Godwin believed that the unlimited power of human reason would defeat every curse of disease, hunger, and even mortality. First, we ask if our choices are simply between Promethean arrogance or Malthusian disillusionment.

CHECKPOINT: MUST WE CHOOSE BETWEEN ARROGANT TECHNO-OPTIMISM OR DISILLUSIONMENT?

In 1933, Newton Baker wrote about the prospects for the League of Nations to put an end to all war. He warned that extreme optimism ignores a grim human history, while extreme pessimism closes the door to even attempting to work on problems and bring moderate improvements in our condition:

The fact is that all efforts to solve human problems are beset by un-identic twin evils. On the one hand, we have the enthusiast who declines to see difficulties, is indifferent to the lessons of history, takes no account of the deep ruts worn by mental habit and, by expecting too much, either in speed or achievement, takes a flight from reality and accomplishes less than the possible. On the other, we have the pessimist, his ranks all too often recruited by disillusioned enthusiasts. Starting with an assumption of human incorrigibility, he soon despairs of any progress in a tough and obdurate world and looks with sour disfavor upon those who would disturb any arrangement which has been found to ease the galling of a burden which it is the inescapable lot of mankind to carry.

(Baker, 1933)

Some scholars believe the League of Nations failed terribly, in part because the United States failed to become a member out of fear that it would compromise its sovereignty, despite the fact that it was the idea of US President Woodrow Wilson, and was one of his famous Fourteen Points that he brought to the Paris Peace Conference that produced the Treaty of Versailles at the end of World War I. World War II spilled the blood of millions of people only a few years after the League began. However, the UN may not have been established without the League's failure, and now wars between countries have substantially decreased. The decline of interstate war is probably not solely because of the UN, *but* UN peacekeeping operations, especially since the collapse of the Soviet Union and the end of the Cold War, have been a central tool in addressing or solving international crises (see for example Kertcher, 2012).

Like war, sustainability presents us with an attitude problem that affects what we permit ourselves to attempt. How do we set our attitude toward sustainability? Put another way, how optimistic or pessimistic should we be about the future, and what justifies the attitude we hold? Humanity has changed the landscape of the world, and invented so many creative solutions to seemingly intractable problems—such as long distance instant communication, flight, space travel, nuclear science, and even tinkering with the genetic code of life itself. At the same time we live in a world that has limits, and just under a billion people, as of this writing, live in chronic hunger and deprivation. That means that about one seventh of the human population is *not* being sustained (Wu et al., 2012). How optimistic should we be that humanity will flourish, for how long, and by what logic is this attitude justified? Unfortunately, we do not have any ecological or social concrete line, short of total nuclear Armageddon, that, if crossed, would cause systemic collapse. We can, however, refer to thinkers of the past, who make proposals.

In the French tradition of memoire, authors present certain facts as they see them to reach an opinion through argument. Here, I use the word memoire loosely, but turning to Desormeaux, Jenson, and Enz (2005), memoire takes multiple forms, including "the documents with which one writes history," but where the memoirist is "not obliged to renounce his passions." I invoke memoire here to identify the rhetorical narrative of these great debates in sustainability so that we may look for more than just facts about sustainability, but instead aim for wisdom. In the history of ideas about sustainability, several giant personalities have unrelentingly haunted the corridors. In memoire, we are reading a perspective and an argument that are neither personal fantasy nor a universal truth that will make sense to everyone. What is true, however, is that these debates are persistent perhaps since the time of Aristotle who speculated about the perfect population of a vibrant city (100,000 citizens).

In sustainability, there is a history of argumentation that speaks from two central perspectives—that of splendid optimism for endurance and flourishing, and that of grim pessimistic ruin. In tracing the spirits of these perspectives, we will understand something important about sustainability—attitude and perception have consequences for what people *do*, and what we do leaves a legacy for what is possible.

THE PERSISTENT GHOST OF MALTHUS AND THE FANTASIES OF GODWIN

How much optimism or romanticism, pessimism or pragmatism should we have about the future, and by what measure should we base our attitude? In Malthus (1766–1834), we have a classical thinker who presents a persistent, if controversial, tenet that populations are limited by resources. For Godwin (1756–1836) and less-so Boserup (1910–99), ecological limits are temporary, and human innovation and even destiny are the true resources that define human continuity and prosperity. First, we will turn to sustainability's Godfather, Malthus.

Malthus is one of the most important thinkers in sustainability studies, but he is a very controversial figure—so much so, that "Malthusian" can be used as something like an insult. The central contribution from Malthus comes from the statement, "Population, when unchecked, increases in a geometrical ratio. Subsistence only increases in an arithmetical ratio," which is the thesis of his primary contribution, *An Essay on the Principle of Population, as it Affects the Future Improvement of Society, with Remarks on the Speculations of Mr. Godwin, M. Condorcet, and Other Writers* (1998), first published in 1798 with six successive editions.

The central tenet of Malthusian theory is that populations grow geometrically, which is at an exponential rate or faster, but that food will—at best—increase only arithmetically, or linearly.

What this means is that population can grow very fast, but he thought food could only grow slowly. In fact, Malthus actually believed that food would decrease over time, because higher populations would need more land to live on and they would do this by colonizing fertile agricultural land. However, he decided to err on the side of caution and say that food would only arithmetically increase. The mismatch of rates in population and food increases will always produce more "men than corn" and induce horrific suffering.

Figure 3.1 shows that, indeed, world population was stable for 1800 years (shaded area), but during the twentieth century has indeed grown exponentially. Thankfully, food production, thanks to industrialization of agriculture in the **Green Revolution**, kept pace with this enormous population explosion.

Petersen (1971) argued that we should distinguish the young and later Malthus. The younger Malthus believed that the population–food trap was inevitable and inescapable, for the poor, and there was little anyone could do about it. The later Malthus began to include the ways in which human action through policies could, only

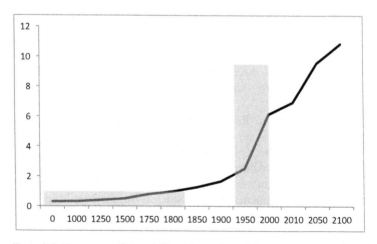

Figure 3.1 Human population (billions) beginning of the common era projected to the year 2100. The horizontal shaded area is a period of stable population under 1 billion, and the vertical shaded area is a period of exponential growth.

Source: Adapted from United Nations Population Division Data (2012)

modestly, relieve this trap. Malthus proposed two checks that would keep population from climbing infinitely. Preventative checks are measures we choose ourselves, such as policy or marrying later in life. Positive checks are reductions in population that are imposed on people. Positive checks on population growth come from the strict carrying capacity or distribution of resources, and when these resources are exhausted, resource scarcity will result in the misery of famine, disease, and violence—perhaps even war:

> Gathering fresh darkness and terror as they rolled on, the congregated bodies at length obscured the sun of Italy and sunk the whole world in universal night. These tremendous effects, so long and so deeply felt throughout the fairest portions of the earth, may be traced to the simple cause of the superior power of population to the means of subsistence.
>
> (Malthus, 1998)

Ultimately, populations will collapse. Then, when times are better and food is more abundant, population will rise again exponentially until it reaches the carrying capacity again, so forth and so on. Further, as the poor population grows, the number of poor workers will increase, reducing worker demand and depressing wages, making the poor even more poor. From this point of view, the poor are in some ways to blame for their own poverty, and Malthus did not think poverty was a result of socioeconomic privilege or opportunity.

Some of the changes Malthus considered later in his thinking came from his debates with William Godwin, a Calvinist minister who was married to Mary Wollstonecraft and father to Mary Shelley. William Godwin's principal work was *Enquiry Concerning Political Justice, and Its Influence on Morals and Happiness* (Godwin, 1798). Godwin founded philosophical anarchism, and in *Enquiry*, he argued that government corrupts human potential by fostering ignorance and dependence through interference. Over time, human knowledge would grow such that our mind would conquer matter, and rules and law would no longer enslave individuals. This laissez-faire, anti-regulatory ideal still persists in those who believe humanity will conquer any problem, including resource scarcities or other sustainability threats.

Godwin believed that human perfection was attainable over time, and that humanity could liberate itself from the organic and inorganic limits of the Earth to realize a divinely prescribed destiny of peace and abundance:

> There will be no war, no crimes, no administration of justice, as it is called, and no government. Beside this, there will be neither disease, anguish, melancholy, nor resentment. Every man will seek, with ineffable ardor, the good of all. Mind will be active and eager, yet never disappointed.
>
> (Godwin, 1798)

Godwin believed the world population would realize permanent happiness and there will be no births and no deaths because our minds would triumph not only over matter, but over death. He assumed that one day our mind would be able conquer the vagaries of nature and morbidity, where, since we "will one day become omnipotent over matter, … why not over the matter of our own bodies?" (Ibid).

Through discussions with Godwin, Malthus modified his theory and came to believe there was an important role for social policy to reduce both misery and population, but mainly these policies should focus on things like education, not aid. That said, Malthus is better known for his opposition to the English Old Poor Law that provided payment to poor laborers with large families as a dole from the local parish. Malthus believed the law provided incentive to add more children to the poor's ranks, adding to overall misery. Therefore, in order to diminish the overall misery, he argued against the Old Poor Law and this aid was eventually ended (Petersen, 1971).

As Malthus matured, he updated *An Essay* six times. The mature Malthus believed that social conditions, such as better pay and free public education for the poor, would provide opportunities and incentives for the poor to have fewer children. If the poor were transferred to the middle class, they would prefer that lifestyle and comfort, and therefore would have fewer children:

> In most countries, among the poorer classes, there appears to be something like a standard of wretchedness, a point below which they will not continue to marry and propagate their species. … The principal circumstances which contribute to raise [this standard] are liberty, security of property, the diffusion of knowledge, and a taste for the

comforts of life. Those which contribute principally to lower it are despotism and ignorance.

(Malthus, 1998)

Consequently, the mature Malthus advocated for broadening political rights and suffrage (Godwin opposed these), free medical care, while Malthus opposed child labor if it were only for the profit of the employer.

The reality is that Malthus' theory on population evolved somewhat over time to show that social policy and theory mattered because the better off the poor were, the less likely they were to have large families—a point well established in international population and family planning circles today. Also, fertility rates decline when women are empowered and more free, not less free (Sen, 2013). Still, Godwin found Malthus too pessimistic and produced revisions of his own argument as a sustained critique of Malthus. Godwin used Biblical passages while he argued that Malthus' theory lacked any empirical validity, among other attacks:

we have not the smallest reason to believe that the population of the earth is in any way more numerous now than it was three thousand years ago.

(Godwin quoted in Petersen, 1971)

The attack on Malthus by Godwin was so ill-conceived and its empirical statements were, even then, so conspicuously wrong that many believed that Godwin's *On Population* did not even warrant a response.

We now turn to a lesser known and later thinker than Malthus or Godwin, Ester Boserup (1910–99). Boserup was a European economist who made several contributions to the economics of development and the role of women in agriculture.

Boserup argued that food will not be a strict limit on population, but that higher populations can lead to more food. Rising populations, *in societies using non-industrialized agriculture*, will begin to move from a long-fallow, or resting, period, using low technology for farming, to a short-fallow period using more tool and labor intensive approaches to raise the productivity of the land. More land will be brought into cultivation, and people will divide up the jobs more systematically. More cultivation, tools, and efficient arrangement of jobs will arise in a developing agricultural market that will intensify the hours spent cultivating.

> The neo-Malthusian school has resuscitated the old idea that population growth must be regarded as a variable dependent mainly on agricultural output. I have reached the conclusion ... that in many cases the output from a given area of land responds more generously to an additional input of labour than assumed by neo-Malthusian authors.
>
> (Boserup, 2005)

Viginia Dean Abernathy, former Editor of the academic journal *Population and Environment*, writes that Boserup was a fascinating and innovative thinker, whose "breakthrough was in seeing that particular technologies do not develop, and are not adopted, in a social or an environmental vacuum" but that changes in technologies have social causes (Abernathy's Introduction to the AldineTransaction edition of Boserup, 2005). Thus, as social conditions like population and food demand increase, technological and agricultural adaptations can result—increases in population cause increases in food production under limited conditions. However, Abernathy also tells us that,

> Boserup's world knows no environmental limit other than arable land. She would be quite at home with certain economists' cornucopian views that scarcity is always a short-term phenomenon because higher prices, signaling scarcity, drive both/either substitution and/or enhanced technology. One can accept almost all of Boserup's conclusions, I think, but this one.

In sustainability debates, there is a persistent polarity between the limits and possibility, endurance and demise represented by these authors.

PERCEPTIONS TO POLICY

Endurance and ruin are argumentative perspectives that have informed policy debates going back 100 years. For example, does aid assist the poor, or does it just allow the poor to increase their ranks and, in the end, add only more misery? Garrett Hardin is often described as a Malthusian. Hardin, in his essay, "Living on a Lifeboat," published in *Bioscience* in 1974, takes a similar position on aid as Malthus did against the Old Poor Law. Hardin, who described himself as an "eco-conservative," argued that if we think of countries as lifeboats, they have a certain limited capacity. If one offers aid to persons sinking around you in a shipwreck to board your lifeboat, then your lifeboat will sink, aiding no one and compounding misery. The ethical thing to do is offer no aid, and observe the limits of your lifeboat; and, like Malthus, this employs a pretty strict Benthamite utilitarianism, that is—what is right is that which simply provides the greatest number of people with the greatest good.

Notice, however, that these questions do not ask how poverty itself is generated. Also, notice that none of this debate specifically is adapted to contemporary times. Boserup's work only is applicable to "early" agriculturalists who are in the first stages of converting land, and "Her analysis is not intended to be relevant where fossil fuel largely replaces human labor, as in the mechanized agriculture of modern societies" (Abernathy's introduction in Boserup, 2005). Similarly, Malthus' formula for food constraints was broken in the beginning of the nineteenth century and especially in the twentieth century, when "A cascade of new farming technologies developed over the two centuries since Malthus wrote his *Essay*—especially synthetic nitrogen fertilizer and improved seed varieties—allowed crop production on existing farmland to skyrocket" (Paarlberg, 2010). Malthus and Boserup only speak to the dynamics of a pre-fossil fuel industrial world capitalist system. Still, the future remains enough of a mystery it is hard to say just how romantic or pragmatic we should be. Perhaps, when there are between 9 and 11 billion people, have used most of the fertile land, and decreased ecosystem function from these technologies, "a Malthusian limit may finally be reached" (Paarlberg, 2010). On the other hand, romantics and cornucopians (see p. 74) alike have an abiding faith that things will work out and we will devise new technologies and new adaptations that look just like the potential Malthus missed.

Today, we see elements of Godwin's optimism in the **Environmental Skepticism Countermovement** that rejects the notion that there are any threats to human sustainability, except for the obstructions that environmentalism poses to industrial progress.

This countermovement is modeled off of the work of Julian Simon. In the 1980s, Julian Simon (1981), a business professor at the University of Maryland, wrote a number of books arguing that people were the "ultimate resource," and that population increased the brain trust of the world and its ability to solve problems. Further, Simon was a staunch advocate of free enterprise and believed that government should not be involved in many environmental efforts; and, he was opposed to environmental groups, whom he thought should be monitored by a business alliance as something of an enemy (Simon, 1999). Simon's optimism eventually was named "cornucopianism" because it imagined the infinite flow of food and unbounded resources in an infinite set of opportunities for humanity—just as the cornucopia fed Zeus an infinite and magical flow of food as he was hidden from the Titans. Simon (1999) even believed that some laws of physics did not apply to economics. This is as unrealistic as Godwin's assertions that eventually all people will become sexless angels who neither experience disease or death.

However, at the same time, human carrying capacity has not been static in the past. Even the most famous modern Malthusian, Garrett Hardin (1995), noted that humanity had *increased* the capacity to support higher populations through the Agricultural and Industrial Revolutions. Critically, Hardin did not think there was any more ecological slack to continue raising populations.

The point is that memoires of hope and warning in their extreme versions are inaccurate, but they both have something to contribute to our thinking. People and societies develop and innovate ways to solve problems that we may not have imagined before; however, the Earth is not infinite in its capacity to support plant life, ice at the poles, water supplies, minerals, space, or the human family.

Importantly, these memoires affect policy, because the assumptions that decision-makers hold can drive endurance or ruin. Paarlberg (2010) notes that British officials took the Malthusian premise that over-population and starvation were inevitable, and therefore allowed the Irish potato famine (1845–49) to kill about a million people, while another million fled, many to the United States. The assumptions of the British led to their decisions and behavior about "doing nothing."

The opposite extreme is just as dangerous because it encourages us to fail to plan or make prudent advance considerations in case the optimism is wrong. Cornucopian extremists deny that there are any problems relevant to sustainability, that people are exempt from the limits of evolution or ecological pressures, and therefore "the term 'carrying capacity' by now has no useful meaning" (Simon and Kahn, 1984). **Human exemptionalism** is such a powerful force that it is considered a paradigm, or a constellation of values, that guide human decision-making. Human exemptionalism is the set of beliefs that humanity is so distinct and special that it is exempt from the rest of the laws of nature, ecological limits, or even evolutionary pressures, including extinction.

Human exemptionalism implies that policy makers do not need to do anything to solve problems like global warming or biodiversity loss if human interests are the only concern. Of interest, the *American Journal of Physical Anthropology* published a three-part series of essays indicating that if global human populations were not limited, Malthusian misery would *continue to worsen* as some 80 percent of the world population lives in "conditions ranging from mild deprivation to severe deficiency" with little indication that this trend will change significantly (Smail, 2002). Smail repeats Hardin's (1995) concern that there is a difference between *optimal* population and *maximum* population—because two variables cannot be maximized at once in any equation (read: you can only *maximize* either wellbeing or numbers of people).

Given the different ways people see the two memoires, one would expect that the published response to Smail would critique his position. Instead, Jeffery McKee responded that human population size and growth statistically explained the vast majority of biodiversity loss; and, that inasmuch as biodiversity is a key to human survival, human population growth will continue to "disrupt ecosystem functioning, thereby threatening our own life-support system and planetary sustainability. It is Malthusian principles writ large" (McKee, 2003). Similar relationships to rising populations and their impacts through fishing, coastal development, and land use are shown to affect 75 percent of the world's coral reefs, and the biodiversity loss therein (Mora et al., 2011).

While it is dangerous to think that hunger and violence are "natural," it is equally dangerous to think that population size and human impacts have no consequence for our current and future welfare. It is also important to realize that simple populations do not have automatic metabolisms, where the larger political economic conditions determine how much more that the minimal metabolism any population group may have—populations always sit inside the larger economic forces that determine how they consume (Swyngedouw and Heynen, 2003). In this way, simple measures of population are never sufficient, but serve only as a baseline minimum estimate for impact because different populations from Dar es Salaam, Tanzania, and Paris, France or Los Angeles, USA will consume the minimum (if they are being sustained adequately, and in Dar es Salaam this is not the case for a large portion) plus what is determined by the political economic context of that population.

In the end, Alan AtKisson (2012) has put our problem a marvelous way—believing Cassandra sets us free. Cassandra was a prophetic woman of Troy during the Trojan War, but she was saddled with the curse that no one would believe her. Ultimately, she predicted the catastrophe of the Trojan horse and invasion among other fates, only to be marginalized (and ultimately raped, kidnapped, and murdered). Sometimes, environmentalists warning of dire problems are caricatured as "Cassandras," forgetting that Cassandra was never wrong, just ignored.

AtKisson argues that acknowledging the real limits of the Spaceship Earth sets us free to begin inspired creativity and problem solving that is far more productive than denial that the problems exist. The only way to prove Cassandra wrong is to take the problems we face with enough gravity to address and solve them. However, solving these problems may take radical changes to the way the world economy and world politics currently operate.

THE BRUNDTLAND REPORT: THE MEMOIRE OF "SUSTAINABLE DEVELOPMENT"

Another central storyline about sustainability is that of sustainable development. The idea itself is contested because it typically means the human economy can and should grow, perhaps indefinitely, while still keeping the Earth's critical ecosystem services intact. It is an optimistic vision more in line with Boserup and weak sustainability than Malthus and is a dominant theme in global environmental politics. The most important proposition for sustainable development comes from the World Commission on Environment and

Development (WCED), known as the Brundtland Commission for its chair and then Prime Minister of Norway, Gro Harlem Brundtland. Brundtland was a formidable figure who was a medical doctor, public health expert, and a former Minister of Environment for Norway—the first environmental minister to become Prime Minister of any country.

The WCED was charged by the United Nations (UN) to formulate a "global agenda for change" in order to "achieve sustainable development by the year 2000 and beyond" (United Nations WCED, 1987). WCED defined "development" as the way people make their lot better in their environment. The Brundtland Commission delivered its report to the UN in August 1987, *Our Common Future*:

> Humanity has the ability to *make development sustainable to ensure it meets the needs of the present without compromising the ability of future generations to meet their own needs.* The concept of sustainable development does imply limits—not absolute limits but limitations imposed by the present state of technology and social organization on environmental resources and by the ability of the biosphere to absorb the effects of human activities.
>
> (United Nations WCED, 1987, emphasis added)

The italicized portion is the most cited definition of sustainability. The notion of sustainable development has a certain allure to those who want to continue the status quo of global economic growth while hoping to protect resources for the future *and* solve world poverty. Figure 3.2 shows the growth of the world's population broken down very roughly between affluent and poor regions, where affluent regions' population is stable, but almost all of the world's population growth is in poor regions and these regions are not changing radically in status away from poverty.

The Commission explicitly noted that the state of poverty in the 1980s was unsustainable since poverty was a systematic source of death and suffering, and, therefore, was contrary to any kind of development, sustainable or otherwise.

The Commission concluded that widespread poverty is no longer inevitable. Poverty is not only an evil in itself, but sustainable development requires meeting the basic needs of all and extending

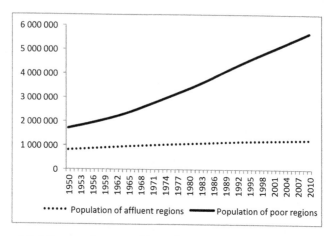

Figure 3.2 Affluent and poor world population 1950–2010
Source: United Nations Population Division Data (2012)

the opportunity to fulfill their aspirations for a better life to everyone (United Nations WCED, 1987).

The Commission called on the world to:

- Grow economically, especially through better technology, to allow the poor to aspire to a better life. They believed the world economy needed to grow *Five to ten times the rate of growth in the 1980s.*
- Ensure there is economic equality and cooperation in the national and international systems so the poor receive their share of the Earth's resources.
- Change "norms and behavior at all levels in the interest of all. The change in attitudes, social values, and in aspirations that the report urges will depend on vast campaigns of education, debate, and public participation."

Through this report, the disparity of resource use between rich and poor countries became a main theme of debate about sustainability. WCED noted that people in affluent countries needed to "adopt life-styles within the planet's ecological means" (United Nations WCED, 1987). Richer countries made up a minority of the world's population, but this minority was using a vast majority of the world's resources, usually on the order of 4:1 or more.

The Earth is one but the world is not. We all depend on one biosphere for sustaining our lives. Yet each community, each country, strives for survival and prosperity with little regard for its impact on others. Some consume the Earth's resources at a rate that would leave little for future generations. Others, many more in number, consume far too little and live with the prospect of hunger, squalor, disease, and early death.

(United Nations WCED, 1987)

The "creative ambiguity" of sustainable development advocated by WCED instigated wide-ranging debate (Kates, Parris, and Leiserowitz, 2005). For example, Hardin suspected sustainable development was only possible through "denying the existence of absolute limits" (Hardin, 1995). Hardin agrees with Donald Mann: "'The concept of sustainable development is little more than a gigantic exercise in self-deception,' because those who advocate sustainable development really mean 'sustainable economic growth.' This, in a world of limits, is 'a thundering oxymoron if ever there was one'" (Mann quoted in Hardin, 1995).

The US National Research Council attempted to divide the issue into what should be "sustained" and what should be "developed" and found that critical life biological and ecological systems should be sustained, while we try to develop better lifestyles in child survival, wealth, and services for people (National Research Council, 1999). However, this does not address the place where trade-offs are made and it does not specifically admit any hard limits to the world. In the end, the ambiguity of WCED allowed for the wealthy countries to enter negotiations at the 1992 United Nations Conference on the Environment and Development (UNCED) unwilling to compromise affluent consumption and poor countries refused to compromise about population growth; both groups blamed each other and agreed essentially to do nothing about these core elements at the heart of the problem structure of sustainability.

Sustainable development is also relevant to the UN Millennium Declaration of 2000, which set out over 60 widely agreed upon international goals along with the eight major Millennium Development Goals. These goals include eradicating extreme poverty and hunger, end gender disparities and inequality in education, and foster "environmental sustainability." The latter goal prompted the Millennium Ecosystem Assessment (MEA), which has found that eradicating poverty cannot be achieved without maintaining critical ecosystem services (Millennium Ecosystem Assessment, 2003). Interestingly, some of the Millennium Development Goals, like cutting in half the amount of people living in grinding poverty by 2015, have been met, while others remain elusive.

In an insightful essay, Hempel (2009) identifies key conceptual problems with the term "sustainable development." He argues the term is ambiguous about whether sustainability refers to simply development of communities, societies, and so on, that simply endure, or is it the "development" or "communality" that is to be sustained? He admits the difficulty of being precise about such a complex idea:

> And, what is it, exactly, that we want to sustain? Human health and well-being? Ecological wholeness? The so-called three Es of sustainability—environmental resilience, economic vitality, and social equity?

Worse, different elements we may want to sustain may, in fact, be contradictory. Governments around the world are almost universally aware of the issue of sustainability, but there is a persistent challenge to square economic growth with social equity and justice and environmental integrity, just as Hardin argued.

Further, Hempel warns that while we may abstractly identify something as "sustainable" in the short run, long-term sustainable development may be cognitively out of reach because the systems involved are too complex and constantly moving.

However, sustainable development might also be a way to "catch the Frisbee" of sustainability. At a finance conference in the summer of 2012, an economist for the Bank of England, Andrew Haldane (2012), pointed out that catching a Frisbee requires us to understand complex physics and other dynamics. Yet, many people not trained in physics, not to mention dogs, catch Frisbees regularly. The reason is because we use heuristics, or rules of thumb, that simplify our options under complex situations, and to avoid cluttering our thoughts. Sustainable development may be theoretically problematic, but on one level, we are forced to live on and make decisions day-to-day. Sustainable development would ask us to simplify the general questions—are we making our situation and the situation of others better? If a decision improves ours and others' material wellbeing, builds ecological strength and integrity, and improves equity, then it may be development that is at least *more* sustainable than actions that damage these things. The most important thinkers on these issues are probably Hunter and Amory Lovins, Paul Hawken (2010), and Eban Goodstein.

Unfortunately, sustainable development does not relieve the contradictions of sustainability raised by Hempel, Mann, or Hardin, though it may provide a window of opportunity to think and discuss these contradictions.

ENDPOINTS

WHAT DO WE KNOW?

We know that if policies are founded on the notion that the poor will "naturally" over-populate, these policies will allow people to die unnecessarily and delay action when aid is necessary. Action, policies, and attitudes make a difference about how we plan for the future, and planning prepares societies for crises, perhaps simply reducing them to surprises. Similarly, if we think that humanity is so well-endowed with genius that we will always be able to surpass limits to the land, this too justifies "no action" and a serious lack of planning.

We also *cannot* know that the future will be like the past, where in the past humanity raised the amount of food it was able to produce significantly to keep up with population increases. We cannot know if we have reached the point where it will be very difficult to get much more out of the land without fundamentally wrecking the ecological cycles and systems that make agriculture possible to begin with.

Also, we know that sustainable development is an ambiguous, contested, and even contradictory discourse. At the same time, this discourse has provided considerations for ecological and justice concerns into national and international policy debates. The Brundtland Report initiated broad international, inter-disciplinary collaborative teams to begin working on a coherent new field of "sustainability science," focused on the endurance of systems, evaluating measures, the resilience of societies, the nature of Earth systems, and the history of ruin using rigorous methods (Bettencourt and Kaur, 2011).

CRITICAL CONSIDERATIONS

What are the most important considerations—ecological limits or human ingenuity—that should determine how optimistic or pessimistic we should be about the future of humanity?

How long into the future should we project sustainability of the human species? How does the time frame we choose—say three generations versus 15 generations—affect how much restraint we should have with the Earth's resources and ecosystems?

What do you think should be sustained and what should be developed if the world is to pursue sustainable development seriously?

What rights or claims do future generations hold over our decisions?

What role do you see for memory of the past in telling the storylines of sustainability?

WHAT DO YOU THINK OF THESE SUSTAINABILITY SOLUTIONS?

What if societies broke sustainability problems into "chewable parts" such as through sectors, and restructured policy to require that each sector, like manufacturing or energy production, always improve material welfare, social equity, and ecological integrity in every activity?

What if, when there are major trade-offs between equity, economics, and ecology in development projects, those most impacted by projects are consulted with "veto power" to stop the project?

WHAT DO YOU THINK OF THE FOLLOWING SYLLOGISM?

Premise A: We do not know what innovations or precisely what ecological limits will be most important in the future for human welfare.

Premise B: Taking no action and failing to plan for future problems forecloses options in the future.

Conclusion: We should plan and build human capacity and take action to protect ecosystems and cycles to keep as many options open as possible for the future.

FURTHER READING

Smail, J. Kenneth. (2002). "Remembering Malthus: A preliminary argument for a significant reduction in global human numbers." *American Journal of Physical Anthropology* 118(3): 292–97. doi: 10.1002/ajpa.10088.

Smail, J. Kenneth. (2003). "Remembering Malthus II: Establishing sustainable population optimums." *American Journal of Physical Anthropology* 122(3): 287–94. doi: 10.1002/ajpa.10255.

Smail, J. Kenneth. (2003). "Remembering Malthus III: Implementing a global population reduction." *American Journal of Physical Anthropology* 122(3): 295–300. doi: 10.1002/ajpa.10341.

McKee, Jeffrey K. (2003). "Reawakening Malthus: Empirical support for the Smail scenario." *American Journal of Physical Anthropology* 122(4): 371–74. doi: 10.1002/ajpa.10401.

Smail provides a series of essays thinking about Malthus' contribution from the perspective of modern anthropology. After a thorough review and application of Malthusian ideas, a response was printed in the same journal by McKee, and the corroboration from McKee offers interesting context for thinking about Malthus' general propositions.

MEASURING SUSTAINABILITY

MAP OF THIS CHAPTER

The storyline in this chapter is that, in order to measure something that is both mutually constructed between objective and subjective boundaries, objective empirical measures are needed to compliment subjective judgments about what sustainability is, what it is made up of, and how it is envisioned and fulfilled. Currently, there are interesting and novel ways that measure sustainability, but since each requires some form of subjective judgment, agreement about which measures to use in any circumstance is needed. None of these measures is complete or indisputable, but some are quite compelling. Some of these methods use vocabularies of empirical change, some use metaphor and heuristic. Specifically, we will examine the ideas and measurements of carrying capacity, the Ecological Footprint, and then the World3 model used in *Limits to Growth*. We will briefly look at sustainability indices and will conclude with a look at the notion of Planetary Boundaries. One theme that clearly emerges across the very different approaches is that current measures consistently show that at the global level, humanity is overshooting and undermining its critical ecological life supports, violating P1. We begin with the story of physicists who had the idea of a Doomsday Clock.

THE DOOMSDAY CLOCK: SEVEN—SIX—FIVE MINUTES TO MIDNIGHT

By 1939, during World War II, it was clear that the Nazis had made important ground toward making an atomic bomb. Elite scientists and recent European émigrés to the US, Albert Einstein and Enrico Fermi, both lobbied then US President Franklin Roosevelt to begin nuclear research given the advances of the Nazis, and this effort was eventually named the "Manhattan Project." As a leading figure in the Manhattan Project, Fermi became the first person to successfully control a nuclear chain reaction in 1942, famously on a squash court in Chicago. Three years later, the US tested the first nuclear weapons in the New Mexico desert, and then dropped nuclear bombs on the Japanese cities of Hiroshima and Nagasaki. World War II was effectively over.

Physicists and engineers that had been working on the Manhattan Project founded the *Bulletin of the Atomic Scientists*. Having observed first-hand what the total destructive power of this new science was and the potential it contained, they were witness to a momentous change in human history because knowledge and innovation had instantly exceeded the political and institutional capacity to manage existential danger. In 1947, the *Bulletin* devised a metaphor to communicate these dangers that nuclear power presented to the entire human race through the Doomsday Clock.

In the metaphor, "midnight" is the twilight of our time on Earth. The clock is seen as ticking both closer or farther away, depending on the gravity and urgency of threats and advances to reducing these threats in any given year.

The *Bulletin of the Atomic Scientists* can hardly be described as a fringe apocalyptic group, but they believe that existential threats, threats to existence, have grown and been reduced since the 1940s. By their estimation, climate change and biotechnology bring us closer to the midnight hour. They include biotechnology as a threat because, while they admit it provides benefits and potential solutions, it also is something that includes serious uncertainties as well as options for designing bioweapons in different forms. They are concerned that scientists could inadvertently create more virulent forms of current pathogens, or even new pathogens, just like when scientists in Australia tried to make a contraceptive out of mousepox, but ended up making the mousepox much more virulent (*Bulletin of the Atomic Scientists*, 2007a). They also describe the effects of climate change, which may be slower than a nuclear explosion, "but over the next three to four decades climate change could cause drastic harm to the habitats upon which human societies depend for survival" (*Bulletin of the Atomic Scientists*, 2007b).

Clearly, the metaphor itself is not empirically grounded. It does however bridge a gap between scientific and public understanding of urgent problems. The metaphor of midnight is immediately under-standable, and the *Bulletin of the Atomic Scientists* is a group that has credibility in the public sphere. Thus, what the metaphor lacks in precision, it makes up for in clarity and authority. Through this chapter it is clear that sustainability studies and sciences has developed multiple vocabularies to communicate the varied threats, but these variations come to consistent conclusions about the challenges that lie ahead.

CHECKPOINT: HOW DO WE KNOW SUSTAINABILITY WHEN WE SEE IT?

Is it really possible to measure sustainability if sustainability is an essentially contested concept? We now understand that there are key uncertainties in First Principles to sustainability. P1 states that we must not undermine critical ecosystem cycles and systems that provide human opportunity and welfare, and this principle is widely agreed upon. Yet, when we attempt to put P1 into practice, the vague idea of sustainability becomes deeply politically contested because it is impossible to know precisely what trade-offs for eco-nomic welfare for ecological integrity and/or social equity will really mean in the future. Also, we are quite simply in the dark about, say, when a fish population will collapse or at what point the Greenland ice sheets will cross a critical melt point threshold, or even if this melt point has already occurred. It is quite possible that the critical point has been crossed in the Arctic. So, we don't know how far we can push Earth systems and cycles, and the points at which Malthus' "positive checks" come into play are fundamentally uncertain, regardless of the measure we use below.

This raises the question of how measures of sustainability should be treated. The depth of uncertainty, and even the types of uncer-tainty (about thresholds or about what other people/groups may do under changing conditions), and the risks at stake have prompted environmental advocates to promote the **Precautionary Principle** (PP). The PP indicates that we should not wait for scientific certainty to take action, and that when making an environmentally risky decision we should err on the side of caution and restraint to protect human health, future ecosystem services like medicines, research, the life of non-humans, and other non-economic values. This principle is especially invoked for decisions that are irreversible, such as biodiversity loss, or introduce risks that are not easily broken down naturally, such as persistent chemicals, and those risks that accumulate and grow as they move through the food chain, these chemicals "bioaccumulate" (O'Riordan and Cameron, 1994; Myers, 1993).

However, environmental advocates should use this notion with purpose because economic growth advocates counter this equation by quite simply reversing what is at stake—when making an economically risky decision they argue defenders of neoliberal capitalism should err on the side of non-interference in the market and development to allow for human health, medical research and treatments, and other values that can be procured through financial resources (Beckerman, 2002). The fault lines of sustainability are not easily reconciled, regardless of what measure is used.

However, remember from Chapter 1 and the discussion of responsible knowledge, where reliable knowledge rarely comes from single sources without review. Claims about sustainability are more reliable when they receive thorough review from others, and the claims of any one approach to measuring sustainability are more valid and reliable if it fits within a growing consilience. Even with fragments of weaker individual pieces of evidence, these can form a larger whole that fits together, and when different approaches provide evidence that converges on similar conclusions, we can treat this consilience with more confidence. This is what we see in the disparate methods of measuring sustainability; there are differences but there is also important concordance. Growing concordance in sustainability science grows out of the Ecological Footprint.

THE ECOLOGICAL FOOTPRINT

There have been several very important attempts to measure sustainability empirically, using quantitative measures. Parris and Kates

(2003) note that there are *more than 5000* attempts to create a quantitative indicator of sustainable development, but none is universally accepted. We will focus on the Ecological Footprint here as one of the most important of these indicators.

The **Ecological Footprint** was co-developed by William Rees and Mathis Wackernagel. Wackernagel now heads up the Global Footprint Network, and there are a number of important resources that can be found at this organization's website (www.footprintnetwork.org).

The Ecological Footprint is one of the few empirical measures relevant to the problem structure of sustainability. In particular, the measure attempts to calculate the amount of productive land and sea needed to satisfy the human needs and consumption, and works specifically within the ecological concept of **carrying capacity**, or the maximum amount of any one population or consumption that can be sustained indefinitely without damaging the integrity of the ecological systems and cycles that the respective population requires to continue (see Box 4.1).

BOX 4.1 CARRYING CAPACITY: A CRITICAL MEASURE

William Rees explains the concept of carrying capacity as the boundary condition for any population of any species: "Ecologists define 'carrying capacity' as the population of a given species that can be supported indefinitely in a given habitat without permanently damaging the ecosystem upon which it depends" (Rees, 1992).

Humans also are constrained by this boundary condition: "For human beings, carrying capacity can be interpreted as the maximum rate of resource consumption and waste discharge that can be sustained indefinitely in a given region without progressively impairing the functional integrity and productivity of relevant ecosystems" (Robert, Daly, Hawken, and Holmberg, 1997).

It is inconceivable that the Earth has an infinite capacity to sustain an infinite number of humans, given the simple mathematics of metabolism discussed in Chapter 2. However, the actual capacity for the global human population is contingent upon many secondary concerns, such as the diversity of consumers and transnational socioeconomic classes and trade, human culture, specific geography, and many other conditions that guide how much a group consumes. Indeed, Hardin notes that over time, humanity has increased the Earth's carrying capacity as we learned to grow more food and use fossil fuels; though, he warns this carrying capacity has now been saturated. Further, he reminds us that we cannot pursue both the optimal welfare for everyone and maximize the total number of people at the same time (Hardin, 1995). If carrying capacity is the level of human population that can be maintained indefinitely *without damaging the integrity of ecological cycles and systems*, then clearly the Earth's carrying capacity for humans has been overwhelmed.

The Ecological Footprint measures the amount of productive land and water space for different areas and groups that range from individuals, countries, and even global humanity. This work has strong scientific grounding, and appears to be reliable and reproducible, while at the same time is one of the only "absolute" empirical measures of sustainability that currently exists. The Footprint is an absolute measure because it measures the total consumption and minimum impact of human consumption over the whole Earth, as opposed to a relative measure that would compare consumption, say, between two countries, and judge sustainability based on the comparison. This measure simply asks, how much productive capacity does the Earth have, and how much of this capacity is the human population using? Importantly, this use can be broken down to see how much one person or one country is consuming.

The Ecological Footprint comes from a strong intellectual legacy that includes the work on carrying capacity, and the work of Vitousek, Ehrlich, Ehrlich, and Matson (1986) in the journal *Bioscience*. Stanford scientists Peter Vitousek, Paul Ehrlich, Anne Ehrlich, and Pamela Matson published work that measured how much of the Earth's **net primary production** (NPP) was either used directly, co-opted, or forgone because of human activities, and found that humanity uses ~40 percent, at the time of publication. That is, as plants take in energy and use what they need for respiration, humans had taken control of ~40 percent of the total energy left over, on a yearly basis by the 1980s. This number includes all of the biological productivity used for human welfare. However, if we only include agriculture and forest harvests in this calculation, Smil (2012) estimates that humans take only ~20 percent of NPP. However, Smil does not find the prospects any more encouraging because in the next 30 to 40 years population is expected to grow 40 percent, doubling or tripling our capture of NPP, meanwhile exhaustion of soils, pollution, and irrigation limits may reduce the NPP that is available for human use. Considering Smil's work, Steven Running (2013) comments that, "One of the foundational principles of biology is that a population cannot grow forever in a finite ecosystem—a progressive system feedback of starvation, predation, and disease limits uncontrolled growth" and as we use NPP unsustainably, a lower estimation of this number only pushes the cycle's timeline farther out into the future, but does not change the basic problem, or the "time it will take humanity to reach a crisis point."

We should remember that NPP is the base, *the foundation*, of the food chain, where autotrophs (organisms that gain energy from the sun-plants) provide the energy foundation for *all other creatures*, including humans. Consuming 20–40 percent of global NPP does not leave much else for the 5–15 million other organisms that live on the planet, and this loss of energy and habitat contributes to the current Great Extinction. Consequently, humanity has left the crumbs for the rest of creation, while putting us closer to a "crisis point" because humanity is clearly consuming and undercutting the vital productivity we rely on for critical life supports. As Smil notes, a growing human population with a growing appetite for living a life of mass consumption will continue to take more and more NPP. However, if we depend on the food chain, we cannot sustainably undermine it for very long, and this conclusion is reinforced by the Ecological Footprint.

The Ecological Footprint goes beyond NPP to account for all "natural capital," or ecological goods and services. The Footprint analysis standardizes the "global hectares" of average of productive land and sea space that is available against what humanity uses. However, because the analysis looks at annual productivity of the hectares, e.g., the reproductive growth rate of a forest, then the rate of use can temporarily exceed the productive supply. This is an example of "ecological **overshoot**."

The Footprint only accounts for six human activities that require ecological space:

1 Growing crops
2 Grazing animals
3 Harvesting timber
4 Marine and freshwater fishing
5 Accommodating the built environment (infrastructure, housing, etc.)
6 Burning fossil fuel.

The activity that takes the most global hectares for human consumption is growing crops. Burning fossil fuels requires land and sea area for sequestration, or assimilating the waste into the land and sea. Wackernagel et al. (2002) figure that 35 percent of anthropogenic ("human caused") CO_2 is absorbed by the ocean, and the rest is sequestered on land in forests and soils through the carbon cycle. However, this 35 percent does not take into account the non-linear kinetics/movement of the carbon cycle, which means the amount of carbon taken up in the ocean and land decreases as more of the sinks are used, thus more CO_2 will stay in the atmosphere as the climate warms (Archer et al., 2009). Thus, the human footprint will increase dramatically over time as CO_2 is added to the atmosphere and requires more time and space to be absorbed. CO_2 most likely has a lifetime of tens of thousands of years in the atmosphere under these conditions (Archer et al., 2009), and the more CO_2 emitted now the longer it will take to be absorbed in the future.

Wackernagel et al. conclude in the 2002 *Proceedings of the National Academy of Sciences* paper that humanity has been consuming more reproductive capacity than the Earth supports, and we have been overshooting our resources globally. However, to understand when overshoot happened and by how much depends on how much land we assume should be left as a buffer for biodiversity, which operates as a source of resilience and stability for all ecosystems (Millennium Ecosystem Assessment, 2005c), not to mention the inherent value of life on Earth. They calculate that the human footprint on the earth was 70 percent of the Earth's capacity by 1961, overshot the Earth's capacity by the 1980s, and grew to 120 percent by 1999, or a 20 percent overshoot. This means that to continue 1999 levels of consumption, we would need the productive capacity of 1.2 Earths, assuming that 10 percent of the land should be conserved for bio-diversity. If we use the Brundtland Report's (World Commission on Environment and Development, 1987) suggestion of 12 percent of the land as a buffer, then humanity overshot the Earth's productive capacity in the early 1970s, putting the 1999 overshoot to 40 percent instead of 20 percent. In addition to traditional consumer societies, now the "BRIC" countries of Brazil, Russia, India, and China have grown in consumption. Globally, "humanity is in ecological overshoot, currently using at least 50% more of nature's goods and services than ecosystems regenerate," and this estimate is conservative but growing rapidly (Rees and Wackernagel, 2013). By 2030 we will need the biological capacity of two Earths. China alone has quadrupled its Ecological Footprint since 1970, and is now only second to the US in its Footprint impact. Should China realize the same per capita impact of the US, China would require the entire regenerative capacity of the Earth (Gaodi et al., 2012). Overall, 80 percent of the countries in the world are ecological "debtors" and use more renewable resources than is available in their countries, relying on the remaining 20 percent of countries, "creditors," to fill the deficit. In the US, consumption has risen from around 5 hectares per person to 8, while the US capacity to regenerate ecological resources has declined in the same time from around 7 to less than 4 hectares per person, and the only way to explain growing consumption is that the US consumes from the creditor ecologies (Borucke et al., 2013).

The growing rate of overshoot is of particular concern since increased incursions into the productive capacity of renewable resources diminishes their future ability, and reduces the menu of choices available for human societies to succeed at P2.

EVALUATION OF THE ECOLOGICAL FOOTPRINT

The Footprint provides no time horizon for when total ecological limits will be reached (how long can we live in overshoot?), nor how specific ecosystem services relate to this overshoot. As an aggregated figure, when polled, some experts believe that the next steps to advance the idea will require more details in static and changing forces within this aggregate; while others wish it would be synchronized specifically with the UN System of Environmental and Economic Accounting (Wiedmann and Barrett, 2010). Several commentators have criticized the assumptions about the use of energy and how much carbon dioxide should be assumed as a baseline in the calculations and consider it "bad economics" and providing little advice that is policy relevant (Fiala, 2008). Likewise, the Footprint is silent on "what to develop," so it provides no guidance on the needs for P2, in part because simply removing or controlling bioproductive land, e.g., through the built environment, does not describe land degradation well enough (Blomqvist et al., 2013). One problem related to this is that cropland, for example, will only be what is allocated, so it is not possible to exceed biocapacity for this measure, and the Footprint appears to roughly measure more of the ecological deficit related to carbon; but, because of the way that grazing land and forests work in the index, it is possible that the Footprint underestimates the overshoot from carbon by 2.3 times (Wiedmann and Barrett, 2010).

Nonetheless, experts do agree that this tool makes for a good bridge between science and the public to communicate the problem of sustainability overall, and it has even led to the development of a specific water footprint (Hoekstra and Mekonnen, 2012) and carbon footprint that are gaining precision and use and that are improving our overall understanding of consumption patterns. The goal of the Footprint is to measure the productive ecology humankind and other sub-groups consume in units of "whole Earths" (annual productive capacity on Earth), and is grounded in the highest quality of scientific measurement possible at this scale, which is why many city governments, for example, use it to assess their self-sufficiency. The Footprint was first published in one of the most reputable scientific journals and while the Footprint has its detractors, it holds high regard amongst most sustainability scholars. The Ecological Footprint appears to only be the beginning of measuring the human consumption of some of Earth's capacities, but right now it continues to be one of the most important efforts to measure P1, and humanity has conclusively violated this principle. Therefore, *without correcting for this overshoot, it is unreasonable to conclude humanity is safely living within the limits of Earth's life supports.*

MODELING SUSTAINABILITY

To understand specific relationships in a complex system, scholars build models. A model is meant to reflect the main aspects of the system to understand the system better. Models are always more simple, and therefore are never 100 percent precise compared to reality because models are meant to draw out the important aspects, not actually recreate reality in its totality. However, models can be valid.

We know a model is valid if it can recreate past similar systems. For example, we know that climate models can accurately recreate past climates in Earth's history given key ingredients (Stone et al., 2009). Models use assumptions, which come from empirical observation and scientific theory, to reconstruct the system to see what happens when parts of the system are changed and manipulated. In climate models, one of the central questions is "what happens to the Earth's climate system when carbon dioxide levels are increased in the atmosphere?" In the models, scientists run different scenarios that include higher or lower increases to understand how the climate system may change. All of these outcomes from the model are based on our assumptions and the key relationships between the variables included. We can test this validity by trying to recreate past climates in the model, and see how well our model recreates those conditions. If it does not match it well, then the relationships in the model are not accurately represented. However, if it matches the past climates fairly well, then the model can be used to build if–then projections about the future—if this, then probably that. Notice this is not a "prediction" because models do not predict the future and they are not crystal balls. The results from models are instead blueprints to make a projection through if–then statements bounded by assumptions of the current relationships, which can change. Prediction implies "this will happen." Models really provide baseline reasoning for relationships, and help us understand what is more or less likely to occur under simplified circumstances.

Several models have been constructed to understand relationships between human society and ecosystem limits, but we will focus on *Limits to Growth* as the most controversial and important.

THE WORLD3 MODEL AND THE LIMITS TO GROWTH

Among the most well-known publications in all sustainability studies is *The Limits To Growth: A Report for the Club of Rome's Project on the Predicament of Mankind* (LTG) (1972) by Donella Meadows, Jorgen Randers, Dennis Meadows, and William Behrens. **LTG** was published directly before the first United Nations conference on the environment in Stockholm. The study eventually sold more than 10 million copies in more than 30 languages, igniting a fierce set of controversies along the way. After the 1972 first edition, the team published two additional editions in *Beyond the Limits* (1992) and finally *Limits to Growth: The 30-Year Update* (2004). Donella Meadows passed away in 2001 in the middle of preparing the 30-year update. The core question of this model is "How may the expanding global population and material economy interact with and adapt to the earth's limited carrying capacity over the coming decades?" (Meadows, Randers, and Meadows, 2004).

LTG used a computer model first developed by Jay Forrester at the Massachusetts Institute of Technology (MIT) to answer this question, commissioned by the Club of Rome. The Club of Rome was founded by Aurelio Peccei and Alexander King, the Director General for Scientific Affairs for the Organisation for Economic Co-operation and Development (OECD) in Paris after the two met in 1967. Peccei, an industrialist and board member for the Italian car company Fiat, wanted to challenge the catastrophic trends that humanity was pursuing, even outside of war. He noted before he died in 1984 that, "if the Club of Rome has any merit, it is that of having been the first to rebel against the suicidal ignorance of the human condition" (The Club of Rome, 2006). The Club of Rome was eventually influential enough to introduce LTG to many heads of state, giving the book a policy outlet few academic studies enjoy. The group has existed, however, mostly in a shroud of mystery. However, Peter Moll was able to study the Club for a few years. Moll learned that the Club is limited to a select 100 people mostly from

Europe, the US, and Japan, and a third from poor countries whose overarching goal is to address the most pressing global and interconnected problems, or "the global problamtique" (Moll, 1993). The Club continues to publish and work on these problems, but it is best known for its publication of LTG.

The LTG model was predicated on identifying the Earth system dynamics of human population growth and the material economy. The model therefore is global and treats humanity as an undifferentiated whole in order to consider the range of possible futures. LTG specifically warns that the model does not predict what will happen, but it does rule out some futures, such as infinite economic and population growth, as unrealistic.

LTG provides some key concepts and theories for sustainability studies. One is the notion of **overshoot**, also used by the Footprint above. Overshoot is the overuse of available resources, but still provides a time for correction before a collapse. After overshoot has occurred, a correction can be made to observe the limit; but, if no correction is made then the system will collapse. One illustrative metaphor they offer is that if we are driving and we overshoot the painted line on the road marking the edge, we need to correct the car, or we will career off of the road. Overshoot occurs due to "(1) rapid change, (2) limits to that change, and (3) errors or delays in perceiving the limits and controlling the change" (Meadows et al., 2004). The rapid change LTG was most concerned about was **exponential growth**, or growth by a fraction of the stock over time at a constant rate, of world human population and world industrial production.

Population and industrial production both produce **throughput** and therefore consume resources and sinks. Throughput is the economic activity of "take, make, and waste" from and into natural systems, consuming ecosystem capacities. In the Earth system, the limits to growth are a complex dynamic between consumption, limits, and feedback loops. A feedback loop is either positive or negative. A positive feedback causes further growth, and negative feedback reduces growth. Both population and industrial **capital** have a "birth" and "death" feedback where people and machines can reproduce themselves, so to speak. Population is fueled by fertility rates and actual births, and mitigated by mortality rates and actual deaths. More steel factories can create more steel factories, and more steel can build more houses fueled by investment rates and is mitigated by depreciation and the average lifetime of the houses or factories. Further, economic welfare affects fertility, births, mortality rates, and deaths. World3 incorporates this factor through the notion of the **demographic transition**—the theory that pre-industrial groups have high fertility and mortality rates and slow population growth. During industrialization, health and nutrition improve, increasing lifespans of the population, spurring a fast population growth rate. As populations become more secure women have fewer children and population growth rates stabilize or decline.

Because of the feedback loops in the World3 model, the model mainly produces non-linear scenarios. They give the example of a non-linear relationship, where people who double their vegetable food caloric intake from 2000 to 4000 may see life expectancy increase 50 percent; while doubling calories again may only add 10 percent life expectancy and any more than that may decrease life expectancy (Meadows et al., 2004). Positive feedbacks in the World3 cause accelerated growth of both world population and economic activity, and make collapse of world population and the economic system a sobering possibility.

HOW DOES THE WORLD3 MODEL WORK?

The World3 model in LTG incorporates the feedback loops of population and capital into source and sink limits of the Earth, and all of the limits can be extended or shortened by technology and adaptations added to the model each time. These limits include cultivated land, land fertility, crop yield, **non-renewable resources**, and the ability to absorb pollution. Each of these measures is global, so in thinking about pollution, the model thinks about the half-life of "pollution" around the world. They underestimate this measure by assuming that all pollution in 1970 had a half-life of one year (if there was no more pollution emitted, half of it would be gone by 1971). Many persistent pollutants have a much longer half-life than this, but it keeps the model from over-estimating the effect of pollution. All of these dynamics, and many more, are entered in the model that affect and are affected by growth processes, limits and delays, and erosion processes (the opposite of growth). In combining all these factors, the *Limits to Growth: The 30-Year Update* produces 11 Scenarios. Each Scenario produces three graphs for 1900–2100 that track the state of the world, material standard of living, and human welfare and footprint.

Each "state of the world" graph maps the trajectory of global average human population, industrial output, food production, resources, and pollution. Each "material standard of living" graph shows global average food production per person, services per person, average life expectancy, and consumption goods available per person. The final graph shows the "trends in human welfare" that include life expectancy, education, and gross domestic product indices, and the human Ecological Footprint. Figure 4.1 shows the trajectory for collapse, if we continue directly on the path of industrial and human population growth.

According to LTG, humankind has exceeded the biophysical limits of the Earth and is consistent with the Ecological Footprint, even though there is much more complexity in the World3 model. While the models produce timelines, LTG argues that these timelines are best understood as relationships, *not actual predicted changes in any one particular year or decade*. In other words, the relationships of all these issues are more important than any precise year or even decade marked in the graphs. That said, even when the model assumes double the known resources in the world or includes technological advances, the models repeatedly show collapse due to overshoot. According to the LTG, it is clear that human population, life expectancy, human welfare, and consumption of goods and services that provide for everyday needs are at risk of collapsing in the twenty-first century. The tendency for the model to produce some type of collapse is very robust, regardless of the adaptation, because the relationships of consumptive feedbacks remain fairly unchanged—even assuming 100 percent more resources in several of the models. Adaptation and decisions affect these trends usually by pushing the grim results out further in time, but unless *consumption of resources*, including food, and the production of pollution are substantially reduced the models indicate the current global world order is not sustainable.

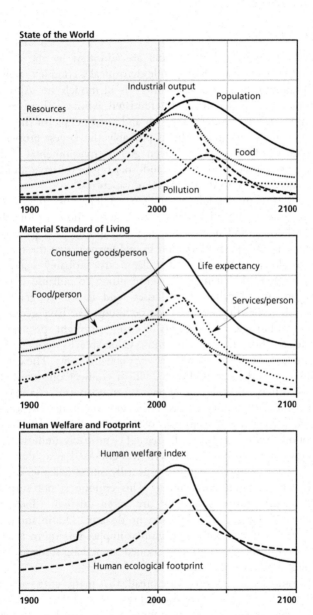

State of the World

Industrial output

Population

Resources

Food

Pollution

1900 2000 2100

Material Standard of Living

Consumer goods/person

Life expectancy

Food/person

Services/person

1900 2000 2100

Human Welfare and Footprint

Human welfare index

Human ecological footprint

1900 2000 2100

Figure 4.1 The *Limits to Growth*, Scenario 1, the standard run, aka business as usual here

EVALUATION OF LTG

Many shortcomings of the model are admitted by the authors. They admit that the model is, while dynamically complex, a serious over-simplification of the real world—as all models are. Also, several things are treated as undifferentiated wholes—there are no geographic distinctions, and the rich and poor people of the world are not treated separately. This is probably the largest problem in the accuracy of LTG because even global environmental changes manifest in regional variations and social groups have different power and capacity to deal with these variations.

Further, the model ignores military capital, does not separate distinct resources or pollutants (except as either short- or long-lived toxic pollution), and neither war nor corruption are a part of the model. What this probably means in real terms is that the relationships of global overshoot and collapse are probably *understated*. Thus, the general dynamics are more prone to collapse than the model presents. Further, since crises tend to be geographically defined, the crises in the model can be assumed to have geographic centers and linkages. The centers of crisis, we might speculate, are likely to be those that are the most vulnerable to ecological and economic problems like extensive poverty and inequality. These centers are linked to other areas though, which connect one area's problems to others, and this can be seen when economic conditions in one area affect other distant areas. The beginning of the downturns in the models may at first seem like local or episodic problems.

It sounds strange to conclude that LTG probably understates the problems of sustainability because it was criticized as a "doom and gloom" book without validity, especially since collapse was not evident in the twentieth century. This critique is not surprising because critics sounding alarms are often labeled "doom and gloom" or "doomsayers" by those who disagree. Doom and gloom is a pejorative phrase with baggage that implies pessimism is always wrong even if the alarm is well-reasoned, something students of sustainability should be made aware.

Criticism of LTG was so widespread, that many critics strangely believe it has been discredited (Lomborg, 2001; Simon and Kahn, 1984). Even the environmental movement had apparently accepted many of the details of criticism against the book (see the documentation of Dobson, 2000). These criticisms argued that the model underestimated resources and human capacity to adapt. However, each model after Scenario 1 assumes that resources are underestimated by 100 percent and are doubled in the model.

Much of this criticism has been unfounded and in some cases an outright misrepresentation of LTG. In his 2008 article, "A Comparison of the Limits to Growth with 30 Years of Reality," Graham Turner observes that many of the criticisms appear to substantially misunderstand the models because many of these critiques made factual errors about LTG itself (Turner, 2008). For example, many critics argued that LTG was incorrect because global society did not collapse by the end of the twentieth century. But the most pessimistic model in LTG, the "standard run" or Scenario 1 depending on the edition used, puts collapse just before the middle of the twenty-first century. Take, for example, Bjørn Lomborg and Oliver Rubin's critique in a 2002 article in *Foreign Policy*:

> But 30 years later, the Club of Rome's most dire forecasts have failed to come true. Vital minerals such as gold, silver, copper, tin, zinc, mercury, lead, tungsten, and oil should have been exhausted by now. But they aren't. Due to an exponential increase in population growth, the world should be facing desperate shortages of arable land and rising food prices. Yet food prices have never been lower. And the world's health should have been undermined by an exponential increase in pollution. People today, however, live longer than ever before, and in Western cities, most pollutants are on the decline, driven down by technological advances and environmental legislation.
>
> (Lomborg and Rubin, 2002)

In fact, most of the models present the case of *rising* life expectancy, industrial output, and consumer goods *until* around the mid-twenty-first century. By 2000, most of the models see slow declines in resources and increases in pollution. Lomborg and Rubin's critique, then, makes it appear they misread or misunderstood the models, but still this position is published in a high profile magazine under the "dustbin of history."

Another standard critique of LTG is that it does not take into account technological advances or market adjustments—but these considerations have a whole chapter dedicated to them in LTG *and* are represented in Scenarios 3–6. In each of these scenarios, collapse is delayed with pollution control technology, larger land yield, more access to non-renewable resources, land erosion protection, and resource efficiency technology but none of these avert collapse in population or human welfare by 2100. To be clear—Scenario 6 does not end in a population collapse, but ends in plunging human welfare due to increasing costs of avoiding hunger. Another critique is that all of the scenarios result in a collapse, which implies that sustainable futures are impossible. However, Scenarios 6, 9, and 10 do not end in population collapse, but these require substantial action in reducing the feedbacks of consumption (and in Scenario 10, it shows what could have happened had the reductions occurred in the 1980s).

Turner then takes the scenarios from the 1974 edition of LTG and compares them against the actual trends after 30 years of historical data and *finds that we are on track for the most pessimistic scenario, the standard run model*, seen in Figure 4.1. Turner further notes that current environmental threats resemble the kinds of things LTG warned about (end of cheap oil, climate change, etc.), and that those models including scenarios, like "total technology" where all the technological advances to preserve food yield and reduce pollution are available and perfectly used, are much too optimistic compared to the historical data. Turner concludes:

> As shown, the observed historical data for 1970–2000 most closely match the simulated results of the LtG "standard run" scenario for almost all the outputs reported; this scenario results in global collapse before the middle of this century.
>
> (Turner, 2008)

Further,

> In addition to the data-based corroboration presented here, contemporary issues such as peak oil, climate change, and food and water security resonate strongly with the feedback dynamics of "overshoot and collapse" displayed in the LtG "standard run" scenario (and similar scenarios). Unless the LtG is invalidated by other scientific research, the data comparison presented here lends support to the conclusion from

the LtG that the global system is on an unsustainable trajectory unless there is substantial and rapid reduction in consumptive behaviour, in combination with technological progress.

(Ibid)

For further corroboration, Joseph Rotmans and Bert de Vries (1997) developed an even more complex model of human health, lifespan, population, availability of water, land, food, energy, and biogeo-chemical cycles—and they also showed overshoot and collapse as one possibility (De Vries and Goudsblom, 2003). All in all, LTG is a groundbreaking measure that helps us understand the global limits to consumption. It appears that the larger warning of LTG is appropriate and accurate, starting from the initial version from the 1970s. Importantly, LTG strikes directly at First Principles and provides critical insight into the intersection of P1 and P2, and Turner rightly indicates that LTG cannot justifiably be ignored. Anecdotally, this author asked a list-serve of professional researchers and instructors of global environmental politics to help identify "errors" they found in LTG, and not a single error was suggested. Also, while the 1972 edition created fierce debate, the latter editions did not elicit much opposition.

Why was this so? In the years between 1972 and now the awareness of environmental effects and problems has changed dramatically. Few people would now openly disregard the environment as one of the most important factors shaping our future ... . *Books like Limits are no longer provocative.*

(Moll, 1993, emphasis added)

INDICES

Moldan, Janoušková, and Hák (2012) write, "The concept of sus-tainable development and its three pillars has evolved from a rather vague and mostly qualitative notion to more precise specifications defined many times over in quantitative terms." Consequently, there have been many efforts to make explicit sustainability indices, or measurable comparisons through compilations of data. Sustainability indices typically aggregate data about water quality, biodiversity, governance, and other factors to summarize sustainability, according to the person/group making the index.

Most scholars or governments making indices agree that the three "Es" of equity, economics, and ecology (or people, planet, prosperity) that make up the so-called Triple Bottom Line (see p. 105) should be represented by reliable data on these elements of sustainability. Since there is no simple measure for overall sustainability, indices are made up of indicators of each part of sustainability. For example, for social equity, we might use the GINI index as an indicator, which measures income inequality by country. For each area with an indicator, the numbers are averaged (typically) to come to one single number that indicates the composite of all the parts of sustainability for that country, city, or corporation. At the end, there is a single number that we can compare Sweden against Zimbabwe to see which is more sustainable, assuming the index is reliable and representative.

One of the more well established indices is the Environmental Performance Index (EPI) (Emerson et al., 2012), published by Yale and Columbia universities in collaboration with the World Economic Forum. The EPI ranks countries by assessing 22 indicators from ten issue areas to assess environmental public health and ecosystem vitality.

However, one problem indices like the EPI face is that of "leakage" because if we measure how clean the water is in Sweden, it is possible that Sweden is importing goods that pollute water, but it is just not polluting Swedish waters. This is still a sustainability problem for Sweden, but if we only have an index on Swedish water quality without capturing this leakage, the index is inaccurate. Another problem is that countries like Sweden and Zimbabwe have entirely different geopolitical conditions that impact sustainability and governments in each country, and indices typically have trouble making sense of these comparisons. One way to deal with this problem is to compare similar country groups, perhaps those of southern Africa against each other.

HEURISTICS AND METAPHORS

Heuristic and metaphorical methods to measure sustainability are more subjective than something like the Ecological Footprint, but they can illustrate important dynamics, threats, and opportunities to sustainability. This section discusses the examples of Planetary Boundaries and the Triple Bottom Line. **Heuristics** are rules of thumb that allow us to draw rough inferences about future probabilities. Sometimes heuristics are called cognitive shortcuts because they are used to make judgments with limited information. This approach has both problems and benefits. The problems are that it is easy to make faulty judgments with unrepresentative heuristics. If a heuristic is not representative of the problems at hand, then it is easy to infer an inappropriate conclusion. On the other hand, it is widely acknowledged that people do not and cannot reasonably expect to use every piece of information to make a good decision, and heuristics help us narrow down what might be important to make better decisions.

Another way to think of a heuristic is as an experimental model for trial and error to promote learning. In this way, perhaps all of the measures for future possibilities for sustainability are heuristic. In fact, Kai Lee (1993) makes the compelling case that adaptive management—trial, error, and correction—are central to sustainability, and this is compatible with the Adaptive Cycle (see Chapter 2).

Thus, the mark of a good heuristic is that it is representative of sustainability First Principles and that it provides an avenue to learn and adjust after mistakes, like correcting for overshoot. Some of these heuristics are grounded in more empirical science, and some are more ad hoc, but each of the heuristics below includes normative (value-based) judgments and makes interesting contributions to thinking about sustainability. In fact, because sustainability includes something about futures, heuristics provide some of the richest literature on sustainability.

PLANETARY BOUNDARIES

In 2009, Rockström et al (2009) provided an empirically grounded heuristic, **Planetary Boundaries**, to answer the question, "What are the non-negotiable planetary preconditions that humanity needs to respect in order to avoid the risk of deleterious or even catastrophic environmental change at continental to global scales?" Their answer is graphically represented in Figure 4.2:

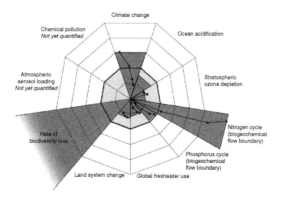

Figure 4.2 Planetary Boundaries

Figure 4.2 indicates that we have potentially exceeded three important planetary boundaries: climate change, biodiversity loss, and the biogeochemical cycle of nitrogen. Table 4.1 lists and explains nine processes that the research could reasonably measure.

Of these nine boundaries, the researchers feel that seven of them can justifiably be quantified, while the research team acknowledges, necessarily so, that these constructs require a combination of scientific and normative judgments. Each of these boundaries has control variables, which are the most important factors that drive us closer to the boundaries threshold points of theoretical changes.

The work builds on the non-linear ideas found in the Adaptive Cycle, as well as the modeling in LTG. As noted above, they believe we have exceeded at least three of the nine Planetary Boundaries: climate change, biodiversity loss, and the biogeochemical cycle of nitrogen. Also, because these changes occur within systems, each change is connected to and interacts with other changes. For example, the principal driver of biodiversity loss is land conversion (in Table 2B "Land-System Change") that eliminates range and habitat for plants and animals. Another example is biodiversity loss caused by nitrogen pollution. Nitrogen fertilizers decrease biodiversity in the ocean when nitrogen empties into deltas because it causes eutrophication, or an over-abundance of algae that consumes the (dissolved) oxygen in the water. The process creates dead zones when enough oxygen is consumed that other plants and animals die or must leave. These dead zones had

Table 4.1 Proposed Planetary Boundaries

Earth system process	Control variable	Threshold avoided or influenced by slow variable	Planetary Boundary (zone of uncertainty)	State of knowledge
Climate change	Atmospheric CO_2 concentration, ppm; Energy imbalance at Earth's surface, W m^{-2}	Loss of polar ice sheets. Regional climate disruptions. Loss of glacial freshwater supplies. Weakening of carbon sinks.	Atmospheric CO_2 concentration: 350 ppm (350–550 ppm). Energy imbalance: +1 W m^{-2} (+1.0–+1.5 W m^{-2}).	1. Ample scientific evidence. 2. Multiple subsystem thresholds. 3. Debate on position of boundary.
Ocean acidification	Carbonate ion concentration, average global surface ocean saturation state with respect to aragonite (Ω_{arag}).	Conversion of coral reefs to algal-dominated systems. Regional elimination of some aragonite- and high-magnesium calcite-forming marine biota. Slow variable affecting marine carbon sink.	Sustain 80% of the pre-industrial aragonite saturation state of mean surface ocean, including natural diel and seasonal variability (80%–70%).	1. Geophysical processes well known. 2. Threshold likely. 3. Boundary position uncertain due to unclear ecosystem response.
Stratospheric ozone depletion	Stratospheric O_3 concentration, DU.	Severe and irreversible UV-B radiation effects on human health and ecosystems.	<5% reduction from pre-industrial level of 290 DU (5%–10%).	1. Ample scientific evidence. 2. Threshold well established. 3. Boundary position implicitly agreed and respected.
Atmospheric aerosol loading	Overall particulate concentration in the atmosphere, on a regional basis.	Disruption of monsoon systems. Human-health effects. Interacts with climate change and freshwater boundaries.	To be determined.	1. Ample scientific evidence. 2. Global threshold behavior unknown. 3. Unable to suggest boundary yet.

Table 4.1 (continued)

Earth system process	Control variable	Threshold avoided or influenced by slow variable	Planetary Boundary (zone of uncertainty)	State of knowledge
Biogeochemical flows: interference with P and N cycles	P: inflow of phosphorus to ocean, increase compared with natural background weathering. N: amount of N_2 removed from atmosphere for human use, Mt N yr^{-1}	P: avoid a major oceanic anoxic event (including regional), with impacts on marine ecosystems. N: slow variable affecting overall resilience of ecosystems via acidification of terrestrial ecosystems and eutrophication of coastal and freshwater systems.	P: < 10× (10× – 100×) N: limit industrial and agricultural fixation of N_2 to 35 Mt N yr^{-1}, which is ~25% of the total amount of N_2 fixed per annum naturally by terrestrial ecosystems (25%–35%).	P: (1) Limited knowledge on ecosystem responses; (2) High probability of threshold but timing is very uncertain; (3) Boundary position highly uncertain. N: (1) Some ecosystem responses known; (2) Acts as a slow variable, existence of global thresholds unknown; (3) Boundary position highly uncertain.
Global freshwater use	Consumptive blue water use, km^3 yr^{-1}	Could affect regional climate patterns (e.g., monsoon behavior). Primarily slow variable affecting moisture feedback, biomass production, carbon uptake by terrestrial systems and reducing biodiversity.	<4000 km^3 yr^{-1} (4000–6000 $km^3 yr^{-1}$)	1. Scientific evidence of ecosystem response but incomplete and fragmented. 2. Slow variable, regional or subsystem thresholds exist. 3. Proposed boundary value is a global aggregate, spatial distribution determines regional thresholds.

Table 4.1 (continued)

Earth system process	Control variable	Threshold avoided or influenced by slow variable	Planetary Boundary (zone of uncertainty)	State of knowledge
Land system change	Percentage of global land cover converted to cropland.	Trigger of irreversible and widespread conversion of biomes to undesired states. Primarily acts as a slow variable affecting carbon storage and resilience via changes in biodiversity and landscape heterogeneity.	15% of global ice-free land surface converted to cropland (15%–20%).	1. Ample scientific evidence of impacts of land-cover change on ecosystems, largely local and regional. 2. Slow variable, global threshold unlikely but regional thresholds likely. 3. Boundary is a global aggregate with high uncertainty, regional distribution of land system change is critical.
Rate of biodiversity loss	Extinction rate, extinctions per million species per year (E/MSY).	Slow variable affecting ecosystem functioning at continental and ocean basin scales. Impact on many other boundaries—C storage, freshwater, N and P cycles, land systems. Massive loss of biodiversity unacceptable for ethical reasons.	<10 E/MSY (10–100 E/MSY)	1. Incomplete knowledge on the role of biodiversity for ecosystem functioning across scales. 2. Thresholds likely at local and regional scales. 3. Boundary position highly uncertain.

Table 4.1 (continued)

Earth system process	Control variable	Threshold avoided or influenced by slow variable	Planetary Boundary (zone of uncertainty)	State of knowledge
Chemical pollution	For example, emissions, concentrations, or effects on ecosystem and Earth system functioning of persistent organic pollutants (POPs), plastics, endocrine disruptors, heavy metals, and nuclear wastes.	Thresholds leading to unacceptable impacts on human health and ecosystem functioning possible but largely unknown. May act as a slow variable undermining resilience and increase risk of crossing other thresholds.	To be determined.	1. Ample scientific evidence on individual chemicals but lacks an aggregate, global-level analysis. 2. Slow variable, large-scale thresholds unknown. 3. Unable to suggest boundary yet.

Source: Rockström, J., W. Steffen, K. Noone, A. Persson, F. S. Chapin, III, E. Lambin, T. M. Lenton, M. Scheffer, C. Folke, H. Schellnhuber, B. Nykvist, C. A. De Wit, T. Hughes, S. van der Leeuw, H. Rodhe, S. Sörlin, P. K. Snyder, R. Costanza, U. Svedin, M. Falkenmark, L. Karlberg, R. W. Corell, V. J. Fabry, J. Hansen, B. Walker, D. Liverman, K. Richardson, P. Crutzen, and J. Foley. 2009. Planetary boundaries: Exploring the safe operating space for humanity. *Ecology and Society* 14(2): 32. [Online] URL: www.ecologyandsociety.org/vol14/iss2/art32/ (accessed 5/28/2014). Used with permission.

not been observed extensively prior to the **Green Revolution** which industrialized agriculture, increasing the use of nitrogen and phosphorus fertilizer that filters into rivers that go to deltas. As these agricultural techniques increased, so did the dead zones; now, there are over 400 of these areas that exist permanently, seasonally, or episodically (Diaz and Rosenberg, 2008). We will now examine the three boundaries Rockström et al. say have been crossed.

CLIMATE CHANGE

The Planetary Boundary for climate change is 350 ppm (parts per million, per volume) of carbon dioxide (CO_2) in the atmosphere. CO_2 is a critical greenhouse gas, and in Earth's deep history, during very warm periods, there was a great deal more CO_2 than now in the atmosphere. The Earth was relatively ice-free until CO_2 was reduced to around 450 ppm $\pm$ 100 ppm. The uncertainty range means that the more conservative estimate would require keeping CO_2 to *350* ppm, a level we have already passed. CO_2 concentrations are accelerating, moving from 0.75 ppm per year in 1960 to >2 ppm per year since 2004. In April 2014 the level was 401 ppm (National Oceanic and Atmospheric Administration, 2014). In 2008, Hansen et al. wrote:

> Decreasing CO_2 was the main cause of a cooling trend that began 50 million years ago, the planet being nearly ice-free until CO_2 fell to 450 ± 100 ppm; barring prompt policy changes, that critical level will be passed, in the opposite direction, within decades. If humanity wishes to preserve a planet similar to that on which civilization developed and to which life on Earth is adapted, paleoclimate evidence and ongoing climate change suggest that CO_2 will need to be reduced from its current 385 ppm to at most 350 ppm, but likely less than that.
>
> (Hansen et al., 2008)

Further,

> Humanity today, collectively, must face the uncomfortable fact that industrial civilization itself has become the principal driver of global climate. If we stay our present course, using fossil fuels to feed a growing appetite for energy-intensive life styles, we will soon leave the climate of the Holocene,

the world of prior human history. The eventual response to doubling pre-industrial atmospheric CO_2 likely would be a nearly ice-free planet, preceded by a period of chaotic change with continually changing shorelines.

(Ibid)

BIODIVERSITY LOSS

The rate of biodiversity loss is the most clearly exceeded Planetary Boundary. Rockström et al. warn that, as indicated by other extinction events, this biodiversity loss will probably come with non-linear and irreversible changes to the Earth system, because biodiversity stabilizes ecosystems and losing biodiversity destabilizes ecosystems (Pereira, Navarro, and Martins, 2012). Each species is adapted and evolved within an ecological niche, and perform various functions. The difference in species function is the functional diversity, and functional diversity is a supporting ecosystems service for other Earth systems and cycles and the other Planetary Boundaries—losing biodiversity reduces overall ecosystem function (Cardinale et al., 2012). For example, pollinators, such as bats and bees, are required for effective reproduction of some plants. If an ecosystem has a low level of functional diversity—few kinds of pollinators for example—it is more vulnerable to catastrophic shifts. As already noted in Chapter 1, the Sixth Great Extinction currently underway and accelerating is one of the most profound global environmental changes. This change has already had an effect on human health and opportunity (Myers and Patz, 2009), not to mention is a profound ethical problem for modernity.

That said, the researchers admit that it is very difficult to say with confidence what rate of biological diversity loss should constitute the boundary. This speaks to the fact that, as a supporting service, biological diversity performs functions that are not well understood and science is unable to put a concrete boundary for human comfort on species loss. As a substitute, the scientists argue that extinction rate can be used until science is able to provide something different. Extinction rates are expressed in terms of "extinctions per **million species-years**," or E/MSY. This means that if there are a million species on the planet, one would go extinct every year; or, if there were one species on the planet it would last a million years, on average. The normal extinction rate is established through the fossil record. The average species exists for 1–10 million years, which means the rate of extinction is 1–0.1 species per million species-years. Currently, the best guess is somewhere between 5–15 million current species on Earth. The extinction rate now is *at least* several hundred times the normal rate, and may be as much as "over 1000 extinctions per million species per year" (Dirzo and Raven, 2003). Scientists estimate the Planetary Boundary for species loss at 10 per million species-years. We are *certainly* above this rate, and very likely above this rate by *several hundred times*. Scientists classifying the current extinction rates suggest that the current period is unlike any of the other periods of mass extinction, and it may end up being the one with the greatest in magnitude and intensity (Şengör et al., 2008).

NITROGEN POLLUTION

Finally, the heuristic indicates we have surpassed the boundary for the biogeochemical process for nitrogen, with the cycle for phosphorus close nearly exceeded. The nitrogen cycle involves nitrogen, N_2 (two nitrogen atoms), moving from the atmosphere, where it, along with O_2, is among the most abundant molecules. Nitrogen moves into the soils and organisms and back to the atmosphere. All plants need nitrogen to grow. But nitrogen is a "limiting factor" for plant growth, where, in other words, plant growth is restricted by the most scarce but necessary ingredient—nitrogen. In order for agricultural crops to grow more they need more nitrogen than is naturally available in the soil. Usable nitrogen is scarce in the soil because it is chemically different than the nitrogen in the atmosphere, which cannot be used by plants, and as more crops are grown in an area, the available (or "fixed") nitrogen for plant growth is reduced because the former plants used it.

Fixed nitrogen is difficult to make. N_2 has three bonds between the two molecules, making it so strong that there are only a few ways N_2 can be broken down for plants, one of which is through lightning. The more common way to break down N_2 is through specific microbes that exist on certain plants—these are nitrogen-fixing plants, like legumes, and they convert N_2 into NH_3, or ammonia. So, for centuries, farmers have planted one crop, like wheat, and then rotated that crop with a nitrogen-fixing crop to help replace some of the depleted nitrogen needed in the soil. However, this process is slow and produces less growth than applying fertilizer. One of the great inventions of the modern period has been the Haber-Bosch process which burns CH_4 (methane gas) to get hydrogen. Then, under heat and pressure, N_2 is added to create NH_3/ammonia. The ammonia is then applied to crops to increase crop yield.

Even though nitrogen is naturally scarce, human modification of the nitrogen cycle now converts more nitrogen from the atmosphere than all natural processes combined (Robertson and Vitousek, 2009). This human-converted nitrogen is washed through the soils and into the waterways, where it is a serious pollutant and the main cause of coastal dead zones noted above (p. 106). This pollution contributes to biodiversity loss, compromises air quality, produces a powerful greenhouse gas (nitrous oxide, or N_2O), and pollutes groundwater. Water with too much nitrate in it can limit the way blood carries oxygen in babies, causing a potentially fatal disease known as "blue baby syndrome." Currently, the United States uses about 12 teragrams (or 12 million metric tons) of nitrogen-based fertilizer for growing crops, and of these 12 teragrams, 10 teragrams are wasted and released into the environment as pollution, so plants are only using one sixth of the fertilizer being used (Robertson and Vitousek, 2009).

Rockström et al. argue that we should think of the Planetary Boundary for nitrogen being like a giant valve where we should limit the amount of reactive nitrogen in the environment. Here they tentatively set the boundary at 25 percent of the current amount (150 million tons globally), but they admit that there is no conclusive way of knowing where the actual boundary for this process is at.

In addition to these three boundaries that humanity has exceeded, the Planetary Boundaries heuristic argues we are close to crossing the boundary of ocean acidification and the phosphorus cycle. As described in Chapter 1, since the Industrial Revolution, ocean pH has decreased by 0.1 units (10 percent) as a result of the ocean absorbing about 30 percent of human-emitted carbon dioxide. Rockström et al. comment that, "This rate of acidification is at least 100 times faster than at any other time in the last 20 million years" (Rockström et al., 2009).

According to the researchers, the boundaries for pollution by synthetic chemicals and atmospheric aerosol loading (air pollution) cannot be quantified. They also argue we are within the bounds for land system change and global use of freshwater.

EVALUATION OF THE PLANETARY BOUNDARIES

Like LTG, Planetary Boundaries works on a global scale. According to the reports, we must solve and respect global limits, or face the imminent loss of the comfortable ecological conditions we have come to enjoy.

Unlike LTG, this heuristic does not include drivers of consumption, and it does not include resources, *per se*. Planetary Boundaries do not say anything about P2, but it does measure attempts to define where the line is for P1 in nine critical areas. So, there is nothing in this heuristic for explaining how to live in the boundaries, nor anything about inequality between rich and poor, geography of the boundaries, or power relations in the human systems, like world politics:

> Because the planetary boundaries approach says nothing about the distribution of affluence and technologies among the human population, a "fortress world," in which there are huge differences in the distribution of wealth, and a much more egalitarian world, with more equitable socioeconomic systems, could equally well satisfy the boundary conditions. These two socioeconomic states, however, would deliver vastly different outcomes for human well-being. Thus, remaining within the planetary boundaries is a necessary—but not sufficient—condition for a bright future for humanity.
>
> (Steffen, Rockström, and Costanza, 2011)

But, as a measure of P1, it corroborates other work, like the LTG and the Adaptive Cycle. Also, one of the more valuable aspects of this heuristic is that it is concerned about finding a workable balance for human operational space and Earth system processes. Because Earth systems and cycles are much more important than simple scarcity, Planetary Boundaries provides a more careful heuristic that pays attention to the *structure* of life supports. Of course, most of these boundaries are defined by substantial uncertainty, and so we have to figure how careful we want to be in order to preserve this comfortable space. Clearly, though, boundaries do exist at some point in these areas, and are made more complicated by the way they each interact with each other, and it seems prudent to mark more careful rather than less careful marks in the sand.

THE TRIPLE BOTTOM LINE: CATEGORICAL EVALUATION OF DECISIONS

The **Triple Bottom Line (TBL)** is a qualitative accounting framework for measuring sustainable progress. Related to TBL is what are also referred to sometimes as the 3Ps – people, profits, and planet, correlating with the three "Es" described in Chapter 2. The three areas of ecology, economy, and equity must all concurrently be in good condition for sustainability to be the bottom line.

Here society refers specifically to social justice and related social goals, all of which were also part of and are consistent with the Brundtland Report (World Commission on Environment and Development, 1987). This means that progress is not just seen through the trade-offs of economic growth and environmental values, but specifically includes the criterion of justice. Because companies and governments have widely adopted this conceptualization of sustainability, it has become widely influential. One of the results has been increased attention to the way companies and governments account for their overall impact in the world, including the civic contributions by corporations. Taking responsibility for the impacts of business, and reducing negative impacts while increasing positive public impacts (not just on those who buy their goods) has been called **corporate social responsibility (CSR)**.

When all areas are healthy and have long-term promise, sustainability writ large is more likely.

The Triple Bottom Line, a term apparently coined by John Elkington, is really a way of operating multi-criteria accounting, instead of allowing economistic concerns to dominate without thinking about ecological and social values. Currently, however, some scholars believe that the mainstream political and economic systems of the world have ignored ecological and social considerations, placing economic ones at the clear forefront. Displacing other values for the sole consideration of economic values is **economism**, and Robert Paehlke has noted that "economism is triumphant" (Paehlke, 2004) in the current globalized world. This would indicate that the TBL is not being met, but the measure itself does not provide us with tools to really evaluate global conditions.

EVALUATING THE TRIPLE BOTTOM LINE

There are many good results from firms working on a strong TBL and thinking seriously about CSR, if it is authentic, and several thinkers have continuously made the case that there are innumerable ways in which making production more ecologically efficient will make profitable sense (Hawken, Lovins, and Lovins, 1999). On the other hand, we must not confuse this framework with sustainability of systems.

Clearly, the TBL does not speak to P1, but rather provides a way to consider multiple criteria when making real-world decisions about what projects include in terms of who and what costs and benefits come with these projects. Also, the framework itself is a vessel to be filled, in that it does not make specific claims about how well the planet, nations, companies, or communities are *actually* progressing toward a firm TBL. Rather, it is a tool for these groups to measure for themselves. The metaphor really is asking for decision-makers at all levels to think about more than economic matters, but also to include often non-monetary costs and benefits that make the world more or less livable. In this way, the framework could be used with integrity by these groups or exploited as a façade, and in most cases it will be very hard to know the difference.

To the extent that the measure speaks to the livability of our decisions, the metaphor provides categorical demands that are important when we are concerned with P2. Some have argued that the Triple Bottom Line is actually a definition of sustainability, but this cannot be true because the decisions made in each issue area could be good ones, but they may not be sustainable in the long term and there is no empirical connection to the limits of a habitable space for humanity or non-humans.

In other words, the Triple Bottom Line is quite valuable in making sure decisions are made by accounting for social and ecological costs/benefits, but the concepts do not specifically address the problem structure of sustainability.

ENDPOINTS

WHAT DO WE KNOW?

There are hundreds of sustainability measures and indicators, but the sample of approaches above all share some common elements. None of these measures indicate that the world is on a path toward sustainability, and most indicate that we have exceeded important limits, from climate change, resources, food, and population. This kind of corroboration across the multiple and diverse approaches adds both gravity and legitimacy to taking true sustainability problems seriously, and since this has not happened on any broad scale, the measures indicate social and economic change is required to make sustainability a prospect.

CRITICAL CONSIDERATIONS

Of those discussed, what are the most compelling measures for sustainability?

If you had to make your own measure for sustainability, what would you include? Omit?

What do you think the measures tell us about the First Principles of Sustainability?

What do you think any one of these measures tell us about making policy for the future?

WHAT DO YOU THINK ABOUT THESE SUSTAINABILITY SOLUTIONS?

1 Every city in the world, just like they create budgets, creates a metabolic budget to understand where the material and energy that the city depends on comes from, how much it is consuming, etc. For every hectare of the city's footprint that extends beyond its borders, the city must restore that much land outside its borders.

2 Every country in the world measures how much they contribute to exceeding Planetary Boundaries, and then makes policy and changes its consumptive behavior to remediate that damage.

3 Individual households are contained to consuming only what comes from their own city, and are encouraged to garden (all

lawns are transformed into productive agricultural space), compost, and recycle so that there is little needed from outside the city and 95 percent of the waste in that household is reused in some way. The materials economy is put into a closed loop, so that landfills are closed down, only reuse, composting, and recycling are available and each building is engineered to collect and purify its own water, manage its own waste and sewage, and produce its own energy from the sun and hydrogen batteries.

WHAT DO YOU THINK OF THE FOLLOWING SYLLOGISM?

Premise A:
Compelling evidence indicates global ecological systems and cycles are being overshot and over-taxed.

Premise B:
The timing of catastrophic shifts and collapse in complex social-ecological systems cannot be precisely predicted.

Conclusion:
The international community should mount a transnational "Manhattan Project" aimed at radical changes to the material economy and civic requirement to limit consumption.

FURTHER READING

Bell, Simon and Stephen Morse. (2008). *Sustainability Indicators: Measuring the Immeasurable?* London: Earthscan. Bell and Morse write a clear-headed explanation of sustainability measures. This book is well regarded and a valuable resource.

Mitchell, G., A. May, and A. McDonald. (1995). "PICABUE: A methodological framework for the development of indicators of sustainable development." *International Journal of Sustainable Development & World Ecology* 2(2): 104–23. doi: 10.1080/13504509509469893. This article describes another approach to measuring sustainability through an index built around quality of human life and ecological integrity, and develops a methodology that can be tailored to specific needs of users, but still be consistent with core principles of sustainability.

Mori, Koichiro and Aris Christodoulou. (2012). "Review of sustainability indices and indicators: Towards a new City Sustainability Index (CSI)."

Environmental Impact Assessment Review 32(1): 94–106. doi: http://dx.doi.org/10.1016/j.eiar.2011.06.001. This article reviews major indices of sustainability, including the Ecological Footprint, Environmental Sustainability Index (no longer produced), Genuine Progress Index and more from the perspective of city sustainability, and is a useful summary of core fault lines and the conceptual difficulties of measuring sustainability.

Klauer, Bernd, Reiner Manstetten, Thomas Petersen, and Johannes Schiller. (2013). "The art of long-term thinking: A bridge between sustainability science and politics." *Ecological Economics* 93: 79–84. doi: http://dx.doi.org/10.1016/j.ecolecon.2013.04.018. This article reinforces the cognitive benefits of using heuristics to bridge sustainability science and decision-making, using ideas going back to the German philosopher Immanuel Kant.

ETHICS, JUSTICE, MORAL ORDER, AND OBLIVION

MAP OF THE CHAPTER

The storyline of this chapter speaks to the way moral order relates to life and death, survival and continuity, and sustainability and oblivion. This chapter will explain a few of the most important ethical debates in sustainability concerning inequality and agency/standing. We will discuss the central problem of ethical values and moral reasoning. The Hopi creation story starts off this chapter by illustrating that ethics may open or close doors of restraint necessary for sustainability, and without this restraint, collapse and death are immanent. Moral reasoning for restraint is a central reason why ethics are instrumental to sustainability, and perhaps are tangible measures of what is "good."

A premise of this book is that justice is a necessary component of sustainability, and this notion of justice should include the broad moral order within which all trends of inequality or exploitation are contextualized. This chapter attempts to demonstrate why justice is so integral to sustainability, and investigates important connections of ethics, justice, and the maintenance of critical Earth systems and cycles. Ethics and justice are lasting criteria for global sustainability, going back hundreds, perhaps thousands of years. Take, for example, the case of the Hopi, whose creation stories indicate that moral and ethical failures lead to critical social failures.

In the 1960s, Frank Waters spent time with 30 elders of the Hopi tribe, who would explain that ethics provide guidance to what is socially permissible and therefore provide limits to what societies allow themselves to do. The Hopi are renowned for their secrecy and for keeping many of their ceremonies shrouded from the eyes of a curious but different dominant Western culture.

The elders broke this legendary silence to tell Waters the basic message that things must change or this world would end. And, what's more, they said they knew this not only from prophecy, but because it had happened before. These stories have been passed down for eons to every Hopi child, in hope that the lessons from the past were remembered. These lessons were of primal importance, relating to the very existence of the human race.

Waters reports that the ancient and thriving Hopi people have lived for millennia in the fragile arid high-desert landscape of what is now northern Arizona. They remember that the First People knew no sickness and were happy.

> Although they were of different colors and spoke different languages, they felt as one and understood one another without talking. It was the same with the birds and the animals. They all suckled at the breast of their Mother Earth, who gave them her milk of grass, seeds, fruit, and corn, and they all felt as one, people and animals.
>
> (Waters, 1963)

This was Tokpela, the First World. However, it did not last.

> There came among them a Lavaihoya, the Talker. He came in the form of a bird called Mochni [like a mocking bird], and the more he kept talking the more he convinced them of their differences between them: the difference between people and animals, and the differences between the people themselves by reason of the color of their skins, their speech, and belief in the plan of the Creator. It was then that animals drew away from people.
>
> (Ibid)

Convinced of their differences and having forgotten the directions to respect the creator, and thereby creation, the deity Sótuknang therefore guided a group of people who still lived by the laws of creation to the underground land of the Ants, who took care of them, while Sótuknang destroyed the First World by fire. Afterward, Sótuknang came to the Ant People, and let the people out, into the Second World, Tokpa. It was not as beautiful or abundant, but it was a great world. After being reminded to respect creation, people spread out and multiplied. However, even though everything they needed was on this world, they wanted more and more. Eventually, the people worshiped the goods more than the sacred elements of the living world, wars ignited, and once again Sótuknang and the other deities decided things had been ruined. Again, the few people that kept the laws of creation were put in safe keeping with the Ant People, and the Twins at the poles of the Earth that kept the proper Earth rotation and balance were told to leave their posts—the Earth spun out of control and froze solid. Afterward, the Twins were ordered back to their posts so the ice would break and the climate would warm; then, Sótuknang released the people from the Ant kiva once again to enter Kuskurza, the Third World. He once again reminded them: "I have saved you so you can be planted again on this new Third World. But you must always remember the two things I am saying to you now. First respect me and one another. And, second, sing in harmony from the tops of the hills. When I do not hear you singing praises to your Creator I will know you have gone back to evil again" (Waters, 1963).

But, violence, greed, and warfare became prevalent again, and water of the world was released, making waves larger than mountains, and the Third World was no more.

Consequently, this is the Fourth World, Túwaqachi. After a long journey, the people of the Fourth World spread out into the lands we now know, with the red, brown, white, and yellow people—all with a mandate for brotherhood and racial unity that the Hopi believe they protect for the future.

Each prior world was lost to human avarice, greed, and hubris. People became alienated from their purpose, and their place in the world, putting self-interest ahead of the web of life and their fellow humans. The elders warned that this Fourth World is on track for a similar fate as the other three unless the vast destruction of the Earth was stopped, and people found their human purpose over again.

Central to these lessons are that our ethical systems and the broad moral order that informs how we treat each other and the rest of the world, and therefore, our ethics appear central to our survival. According to the Hopi, ethical failures were the root causes of prior social collapse.

CHECKPOINT: DO ETHICS *REALLY* MATTER TO SUSTAINABILITY?

A hard-nosed realist might think that ethics are a nice luxury, but are ethical social rules really necessary in order to maintain that society? Do we include something like justice in the requirements for sustainability just to sound interested in fairness? The answer is that ethics, moral order, and justice are central to sustainability because they relate the cause of sustainability problems, how well we will be able to address sustainability problems, and these concerns relate to life and death now and in the future.

For example, ethical failures allow for systemic corruptions that put the public in danger, and as such systematic ethical failures are a key feature of Normative Failures introduced in Chapter 2.

William Ophuls laments:

> Besotted with hubris, we cherish the delusion that we can overpower nature and engineer our way out of the crisis [of ecological destruction]. We are not yet ready to admit that the destruction of nature is the consequence not of policy errors that can be remedied by smarter management, better technology, and stricter regulation but rather of *a catastrophic moral failure* that demands a radical shift in consciousness.
>
> (Ophuls, 2011, emphasis added)

Ophuls argues that the only way to preserve humanity from certain extinction is to create a moral order "in the name of some higher end than continual material gratification," based on true laws governed by the dynamics of nature itself' (Ibid). Ophuls remembers the suggestions of the Ancient Roman philosopher Cicero, who believed that such a moral order "written on the tablets of eternity," and that "True law is right reason in agreement with Nature ... valid for all nations, and for all times" (Cicero quoted in Ophuls, 2011). This kind of natural law is consistent with the laws of the Hopi, explained above, and violation of such natural law to Cicero, Ophuls, and the Hopi elders invites social oblivion.

Ophuls makes the case that ecology, through the self-organizing, dynamic limits and balance of nature, indicates that ethics consistent with ecological dynamics are consistent with life, justice, and long-term continuity that would promote a worthy civilization. "Both the wisdom and the ethic follow from the ecological facts of life: natural limits, balance, and interrelationships necessarily entail human humility, moderation, and connection" (Ophuls, 2011). Not everyone will agree that there is such a natural law, because "natural law" has been used in the past to substitute for rules that empires wanted in place without being questioned. However, scholars do agree that the ethical expectations of a society play important roles to limiting resource exhaustion as well as social inequalities. For Ophuls, restraint that stems from a strong moral order, and not a tangle of laws, is a key feature of a just and sustainable society. He argues these limits are central to sustainability because such a moral order across all nations and all times would lead to the maintenance of the homeostatic balance of life on Earth, which is the basis for all other complexity and development.

Others argue that sustainability *is* the maintenance of justice that allows others to realize their rights in both present and future generations (Dower, 2004). Eroding critical Earth life support systems forecloses on options now and in the future, particularly when current generations disrupt climatic conditions and irreversibly eliminate biodiversity. If ecological systems are the base of all material living conditions, strong ecological systems are required for others to fully realize important rights, such as subsistence and security. In this way, thinkers like Dower might argue that whether ethics is necessary for sustainability is not the right question because sustainability is centrally about ethics, and justice of not damaging the conditions for others to fulfill their inalienable rights of being human. Even if we say that we are only responsible for not hurting others, such as through a libertarian ethic with little obligation beyond that, then we cannot justly participate in economic and political systems that harm sustainability because this harms others, especially the poor, who cannot then realize their most basic human rights. An important aspect of how justice is observed, is the way that power is organized, and this raises the question of inequality and political power as it relates to sustainability.

INEQUALITY AND JUSTICE

Globalization advocates argue that globally connected markets have improved growth and global efficiency, allowing for more revenue to flow to countries who exploit their comparative advantage. All of this allows for more food, energy, and consumer goods to reach international markets and higher levels of consumption for those who can afford it. Globalization has also come with some costs. Saith (2011) writes that globalization has also come with "toxic" inequality where, "The richest 1 per cent was found to own 40 per cent of global assets, the richest 2 per cent accounting for one half of global wealth, and the top 10 per cent owning 85 per cent; at the other end, the bottom half of the world's population owned just 1 per cent of global wealth." Saith argues that this inequality helps drive international social and financial crisis because this kind of deep inequality subverts **institutions** in charge of regulation and accountability, corrodes social norms, and undermines democracy. At this point in human history, every other person on Earth lives in severe poverty, while the most wealthy people occupy a very small transnational elite class (Robinson, 2012), and this inequality distributes and affects **life chances**. Life chances are the opportunities an individual has to improve the quality and duration of her/his life. The concept is important to sustainable development because individuals and households with fewer life chances live with more death and disease (Boulanger, 2011). For example, around 60 percent of the world population, over 3.7 billion people, is malnourished; and, even if the *proportion* of malnutrition has gone down, there are more people malnourished now than ever before in human history (Pimentel et al., 2005), and this malnutrition decreases life chances through poor health, including maternal–child life chances.

Inequality also deeply affects who has access to energy. "For the underserved, the best options for reliable energy are decentralized, clean, and renewable options, such as solar power, biogas, etc." but those most in need cannot afford these options so they turn to dirty fuels that cause terrible indoor pollution (Hande, 2007). This indoor air pollution from fuels like wood, kerosene, charcoal, crop residues like stems and stalks of harvested plants, and dung cause respiratory diseases that likely kill over a million children a year (Ezzati and Kammen, 2002; Alam et al., 2012), and is also associated with low birth weight that has lifelong impacts, and stillbirths (Pope et al., 2010). To bring solar energy to rural and poor communities, Hande and others (Polak, 2008) argue that aid is not the answer because it causes dependencies and does not solve the problem of poverty, which is a lack of income-generating opportunities. To raise life expectancy to 75 years for the average person and lower infant mortality rates to less than 1 percent of the babies per 1000 born, each average person needs to consume about 50–70 gigajoules (GJ) of energy per year (Smil, 2000). The source can vary but this is the equivalent of about 1.6 tons of oil per person. This level of energy consumption is also affiliate with literacy rates over 90 percent and widespread access to education. However, the average person in the United States consumes about 300 GJ per person per year, or the equivalent of 8 tons of oil per person.

Death and disease, as well as medical care, good nutrition, and important opportunities, do not spread themselves evenly, but are organized by society through political and economic rules, historical prejudice, and even geography, and deep inequality will mean a difference in life and death though life chances.

In addition to affecting how much deprivation people live with, justice and fairness impact how well we live with ecological systems. In a landmark study about the role of inequality in sustainability, Andersson and Agrawal (2011) studied more than 200 forest areas where people relied on the forest but were not owners of the forest. They found that inequality within and between groups hurt the sustainability of the forest. Every increase in inequality decreased the wellbeing of the forest. Importantly, institutions arbitrate the way inequality affects the forest conditions and these institutions make a difference to sustainability by restricting inequality.

Yet, some thinkers like Ophuls, believe that large groups of people will always be led by a minority, and some inequality in itself is not necessarily bad. What is important is the form of inequality. Few people suggest that total equality of status, income, opportunity, etc. is a good or a pragmatic way to organize sustainable societies. Making everything absolutely equal, so-called leveling, would mean there is no leadership, genius, or status that comes legitimately from skills or experience, and there is little incentive to move beyond mediocrity because any exceptional intellect or insight would be suppressed. In a large society, total equality could translate to mob rule. On the other hand, a plutocratic inequality creates disempowered castes or classes of people who have little ability to participate in rule-making, or positively change the conditions of their lives. It is more difficult for a disempowered group to obtain redress for grievances. If there is a lot of inequality, there is less accountability for decisions required for First Principles. Further, Vollan and Ostrom (2010) indicate that institutions are required for solving sustainability problems, and if people will not participate in institutions they perceive to be unjust (Rawls, 1971), then *just* institutions are required for sustainability.

Finally, and perhaps most importantly, the kind and depth of justice in our societies reflects who we are and who we want to be—our vision. Our vision indicates what we value and therefore are willing to support, invest in, sacrifice for, and work towards. Donella Meadows, co-author of *Limits to Growth*, argued that we need to have a vision of a sustainable world to bring it about, and that the vision needs to represent something of value and the vision needs to be responsible: "There does need to be a responsibility in vision. I can envision climbing to the top of a tree and flying off. I may really want to do that but my rationality and my knowledge of how the world works tells me that's not a responsible vision" (Meadows, 2012).

Further, she believes that the vision of simple growth (in income, building, technology, etc.) has crowded out other more sustainable alternatives. Meadows' vision of a sustainable world follows:

> An alternative, sustainable world is, of course, where resource regeneration is at least as great as resource depletion. It's a world where emissions are no greater than the ability of the planet to absorb and process those emissions. Of course it's a world where the population is stable or maybe even decreasing; where prices internalize all costs; a place where no one is hungry or desperately poor; a place where there is

true enduring democracy. These are some of the things I have in my part of the vision. But most of those are physically necessary or socially necessary parts of the vision. They form the responsible structure that you know has got to be part of the vision. But, then, what else? What more? What would make this a world that would make you excited to get up in the morning and go to work in?

(Meadows, 2012)

She notes that our visions do not need to be immediately achievable, and they may even seem impossible at first. However, as we share our vision with others, possibilities emerge.

ETHICAL SYSTEMS AND SUSTAINABILITY

What ethical approaches are most consistent with P2-type sustainability demands? Ethical philosophers tend to divide approaches among three traditions important for our consideration: utilitarian, deontological, and virtue ethics.

Utilitarian theory, popularized by John Stuart Mill, argues that something is right if it brings more pleasure or less pain to more people than the alternatives. Deontology, popularized by Immanuel Kant, argues something is right if it is categorically right, regardless of outcomes, and people have a duty to act consistently with what is categorically right. Virtue theory was first brought to us by the Ancient Chinese, Plato, and Aristotle. Virtue ethics argues that something is right if that action "respond[s] well to a morally salient fact about the world" (Sandler, 2013). In other words, what is right is contingent on the context of the situation. Here, we can only provide a brief sketch of the overall reasoning, but we can show that philosophers tend to think that utilitarian ethics and political-economic systems that use this ethic are much less compatible with the normative demands of sustainability. However, no ethical system is without its trade-offs and problems for sustainability.

Utilitarian ethics are usually deemed to provide inadequate injunctions and limits to present-day consumption for sustainability (O'Hara, 1998) for multiple reasons, but mostly because any pleasure and pain is equal to any other, and easily traded off for future gains. Since mainstream economics is based on utilitarian ethics, sustainability scholars tend to see a contradiction between sustainability and the mainstream economic ethics; mainstream economic ethics are often deemed inadequate and unsustainable (Dresner, 2008). Utilitarian philosophers argue that what is good is what brings the most pleasure and the least pain to the most people. One exception to this trend may be the utilitarianism of Gifford Pinchot, the first US national forester active during the Progressive Era of Theodore Roosevelt. Pinchot extended the ethic of his charge over forests to be to provide the greatest good, for the greatest number, *for the longest time*. Pinchot believed that it was his job to protect the forests from being raided for short-term present-day needs, but as a national estate the forests were a store of wealth for the nation into the future and the nation's posterity depended on his good stewardship. He argued that the reason the US was the most prosperous nation of his time was because prior generations bequeathed enormous natural wealth, and that this natural wealth was the basis for all future welfare. He elaborated his reasons for preserving these riches:

> Unless we do, those who come after us will have to pay the price of misery, degradation, and failure for the progress and prosperity of our day. When the natural resources of any nation become exhausted, disaster and decay in every department of national life follow as a matter of course. Therefore the conservation of natural resources is the basis, and the only permanent basis, of national success.
>
> (Pinchot, 1910)

The extension of time to Pinchot's utilitarianism appears to give him moral reason to preserve the nation's natural capital for future generations for future welfare, but without the time factor it is unlikely the reasoning would extend thusly. Like other utilitarian approaches, though, his goals were expressly human in nature, where his practice of silvaculture—cultivating forests as crops—was meant to provide enough wood for every American to build a house, and the forest's value did not extend much farther than that. Thus, if there was a substitute for wood to build houses, and the nation did not need the timber, his stated goals would probably be much less protective of the forests.

Deontological ethics in sustainability tend to stress the duty to protect and provide current and future human welfare, as in the human development model. The human development approach assumes a universal set of rights for every individual on the planet, regardless of gender or ethnic group or class. This Kantian approach is advocated by development scholars Sudhir Anand and Amartya Sen (2000). Anand and Sen explain that the universalist perspective argues that we must be impartial, and offer no privilege to the "right group" or the "right gender," etc. but universally, there is, "the recognition of a shared claim of all to the basic capability to lead worthwhile lives" (Ibid). This obligation is to generally pass on to posterity the ability to be well. This does not engender a responsibility to pass on any specific thing, such as a specific forest or water source passed on because some resources are substitutable, but rather the overall capacity to live well, which comes from capital, or assets that create income and capital can be ecological, material, or human. Naturally, it makes no sense to say we are obligated to make sure the future can live well, and then be unconcerned with today's poor citizens, where "Sustaining deprivation cannot be our goal, nor should we deny the less privileged today the attention that we bestow on generations in the future" (Sen, 2013). Thus, a universalist ethic for sustainability requires we invest in education, health care, and nutrition for the poor today as both an end in itself and as a way to build more productive societies that will build stronger future generations as well.

Importantly, the universalist ethic assumes moral equality of all individuals, where every person is worth the same regardless of different capabilities. Under these circumstances, what kinds of inequality are acceptable? Here, political philosopher John Rawls (1971) made the case for what he called the **difference principle**. He argued that if a rational person did not know what kinds of privileges or obstacles s/he had in life and therefore s/he did not know what kinds of resources—rights, wealth, opportunities, etc.—they possessed either, that person would agree to rules that only allowed for inequality when it benefited the least well off. One problem that this approach has is that it indicates what each individual should have, but not what kinds of things should be restrained. Further, Kantian ethics in general have been criticized for being overly concerned with individuals and less so about communities or relationships epitomized in feminist ethics of care (Gilligan, 1995) or many indigenous ethics that focus on both human and non-human communities (Stewart-Harawira, 2012). This last criticism is indicative that ethical systems make sense to specific ways of being in the world, or **ontologies**, and different ontologies will have different parameters for how to sustain that type of being and identity across time. Ontologies also often come with presumptions about what the overall human purpose is, who and what count as other beings, and what the right relationship is with non-humans.

Whereas utilitarian and deontological ethics are typically anthropocentric, virtue ethics tend to recognize the importance of the human context among other non-humans because its goal is to guide human action toward excellence and flourishing of the whole person. Those actions that promote flourishing, like contemplation or simplicity, build a kind of person and society that is worthwhile. Virtue ethics, then, also is different from the above two others because it identifies something to sustain other than needs. Hull (2005) describes environmental virtue ethics as a form of good ecological citizenship; and, "the core values embraced by a person who possesses this excellence begin with the recognition that we are all plain citizens of our planet, that we for our own physical, intellectual, and moral benefit share it with other forms of life." Decisions to develop a natural resource then must meet important criteria—is this an unavoidable and necessary development for flourishing? Is there an alternative? If it is chosen, are there ways to minimize, rebuild, and improve the natural area afterward? Hull notes wealth is merely a means to an end, and shameless pursuit of pleasure and self-indulgence is *wrong* compared with responsibility, prudence, moderation, simplicity, and other virtues that bring a person closer to human excellence. The same calculation can be made for societies that encourage degrees of ecological virtue and vice. Others have noted that virtue demands that each person is responsible for their impact on others, and Dobson (2003) has argued that we are responsible for ensuring that our Ecological Footprint (Chapter 4) does not harm others, anywhere on Earth. Several problems also riddle virtue ethics, namely the list of character traits that would specifically embody excellence and thriving, and it is unclear where the process of determining virtue comes from— is it inherent to a person, or does it come from social rules or cultural contexts? This relativism sets up problems for virtue ethics about how to judge behavior or people who are not in the same culture.

AGENCY AND STANDING

The above discussion indicates that debates about any moral order must be clear about "who counts." This section discusses the way that different ethical systems have privileged and marginalized specific agents and how this affects sustainability.

Ethical reasoning is about deciding what is right, and in sustainability we are very concerned with environmental ethics that debates what is right for humans, non-humans, and landscapes. In Mill's (1974) utilitarian theory, if something is not sentient, it cannot have any more or less pain, and would matter much less than an agent who presumably can feel pleasure/pain, and for Mill this was only humans. One of Kant's more famous declarations was that no person should ever be treated only as a means to an end, as a tool with mere instrumental value, but should be thought of as an end in themselves, having inherent value. This demand makes the universalism for human development a powerful responsibility. But, what of non-humans? For Kant, humanity had an duty to treat animals well because it served the duty to ourselves to appreciate beauty (love of something not useful), but not necessarily for the inherent value of the animals themselves (Korsgaard, 2004). Consequently, Kant's deontology does not hold animals as an agent in themselves, and what people who are moral agents can do with animals is more open than with what they can do to other people.

Different ontological positions assert what kind of entity has **moral standing** and **agency**. To have agency means to have a will. Any entity understood to have agency also then has moral standing, or must be considered to matter. Jenni argues that, "The central question in Western environmental ethics has been that of what has moral standing: what entities deserve direct consideration in moral decision-making?" (Jenni, 2005). Within this Western tradition, only humanity has mattered in both religious and secular circles because the Western tradition sees humans made in the image of God or to have reasoning. "Hence in both branches of Western thought, humans are at the center of the moral universe," (Ibid) and this is captured in the term **anthropocentrism**. Indeed, Peter Hay (2002) has argued that if there is one thing that sets Western thought apart from other traditions, it is the unequivocal support for anthropocentrism.

Anthropocentrism places humans in a special moral category superior to animals, plants, and certainly inanimate landscapes like mountains, and is consistent with the Human Exemptionalist Paradigm (HEP) (Dunlap, 2002) typically based on proposed special human abilities such as sentience, self-awareness, rationality, memory, and language.

Since the 1970s, animal rights activists have challenged the HEP, in order to grant animals more moral standing and protection. Since that time, cognitive and animal scientists have shown that invertebrates all the way to mollusks have the biological receptors, the neurological systems, and the behavioral responses that are consistent and necessary for human pain; and many animals like dolphins and monkeys show evidence of self-awareness, memory, and language (Jones, 2012). While these conditions may be stronger for humans based on the limited understanding we can have about how other species think and feel, clearly there is evidence to suggest that the exemptionalism of humans has been extremely over-stated. Still, there continue to be advocates of the HEP, such as the Center on Human Exceptionalism at the conservative think tank The Discovery Institute, which it says "counters pseudo-scientific attacks on human dignity by defending the unique dignity of persons, what we call human exceptionalism, in health care policy and practice, environmental stewardship, and scientific research" in their pursuit of advancing free markets and promoting the "theistic foundations of the West" (Discovery Institute, 2012).

Capacities, like reason or sentience may not matter though. Virtue ethics argues that our ethical character is determined not just by how we treat those who are *like us* but also those who are different than us—"others." It is a well-worn psychological concept that when groups form, the group we see ourselves in is the "in-group" and others not in this group are the "out-group." In classical psychological studies, like the Stanford prison experiment by Zimbardo (who saw regular people descend into sadists when put into known false prison guard–prisoner roles), and throughout historical events like the Holocaust, people have been fairly easily manipulated to perpetrate horrific acts upon those they have deemed to be in the out-group (Navarrete et al., 2012).

The essential point to all of this is that moral standing confers recognition—if an entity has "enough" moral standing it cannot simply be disposed of like a tool with only instrumental value. Entities with enough moral standing have inherent value, or value that is not dependent on usefulness to anyone else, but that which comes with full agency. Anything that does not have enough moral standing, then, is open for disposal and what is open for disposal can be used and depleted. In Western culture, there are few exceptions to the statement that only citizens and semi-citizens have inherent value, and everything else is property *and* can be disposed of without any more regard than to the rules regarding that kind of property. For example, Western cultures typically have different rules about the treatment of dogs and piñatas, but in the end the dog is still property in Western law. The problem this presents is two-fold. The first problem is pragmatic, the second is ethical.

The pragmatic effect is that a lack of recognition for non-human others allows for their disposal without much accountability. For example, "the moral status of animals as reflected in almost all— even the most progressive—welfare policy is far behind, is ignorant of, or disregards our current and best science on animal sentience and cognition" (Jones, 2012). This disregard for animals in our institutions like laws that govern raising animals for industrial meat production fail to reign in abuse. According to Robert Jones, factory farms produce most of the 1.02 billion cattle and 1.2 billion pigs per year for food and these animals may live horrific lives: "Cattle raised for beef are castrated, dehorned, and branded, all without anesthesia or analgesics. At slaughter, improper stunning techniques can cause some cattle to be hoisted upside down by their hind legs and dismembered while fully conscious" (Ibid). These industrial conditions create manure lagoons that severely pollute waterways, place enormous demand on grain yield, and pose disease threats because the concentrated groups of animals create perfect epidemic conditions. To avoid disease, these facilities prophylactically feed more antibiotics to the animals than are used in all of human medicine, and this creates stronger pathogens immune to the antibiotics, and, "Antibiotic-resistant Salmonella, Campylobacter and Escherichia coli strains that are pathogenic to humans are increasingly common in poultry or beef produced in large-scale operations" (Cassman, Matson, Naylor, and Polasky, 2002). There is a consensus in agricultural studies that modern agriculture has risen to the challenge of feeding more people, but few scholars believe this is sustainable given the costs and contradictions, such as the erosion of soil needed to grow more crops (Pimentel, 2011). There is a connection between the disregard for other animals, let alone biodiversity, and the sustainability problem we face in feeding the world.

The second problem is that disregard for non-humans is not virtuous. There are at least two variations of anthropocentrism which are based on very different moral reasoning. Most people in the Western world hold a **humanistic anthropocentrism** where ecosystems with their plants and animals are valuable and should be relatively protected because they are resources that may someday help another person (Dietz, Fitzgerald, and Shwom, 2005). This view would support protecting coral reefs because we may find answers for cancer treatment from coral organisms but not necessarily because coral deserves to exist apart from its use for humans. In this case, future humans have moral standing, not the coral.

More extreme is **deep anthropocentrism**. This view argues that protecting nature harms humans because it keeps people from gaining resources from nature, and that there is no ethical value to nature at all. The best example of this is in Peter Huber's *Hard Green: Saving the Environment from the Environmentalists, A Conservative Manifesto*:

> At this point in history, the second vision is a lot more likely than the first. We can go it alone. We need energy, nothing more, and we know how to get it from many more places than plants do. We don't need the forest for medicine; as often as not, we need medicine to protect us from what emerges by blind chance from the forest. We don't need other forms of life to maintain a breathable balance of gas in the atmosphere or a temperate climate. We don't need redwoods and whales at all, not for the ordinary life at least, no more than we need Plato, Beethoven or the stars in the firmament of heaven. Cut down the last redwood for chopsticks, harpoon the last blue whale for sushi, and the additional mouths fed will nourish additional human brains, which will soon invent ways to replace blubber with olestra and pine with plastic. Humanity can survive just fine in a planet-covering crypt of concrete and computers.

(Huber, 1999)

This ethical perspective is consistent with and found in environmental skepticism and climate denial, the social countermovements that argue global environmental problems like global warming are not real or important (Jacques, 2009). Huber indicates that this ethical position then allows us to "harpoon the last blue whale for sushi" because the largest creature to ever grace the Earth has no moral standing *and* Huber argues that to interfere with people exploiting nature for nature's sake, such as regulating resource extraction, is *immoral* because such interference harms people. Deep anthropocentrism is consistent with and is inherently integrated with the HEP which also argues that the only moral agents in the world are humans. HEP proponent Wesley J. Smith (2012) argues, "Moral agency is inherent and exclusive to human *nature*, meaning it is possessed by the entire species, not just individuals who happen to possess rational capacities." Smith argues that because only humans have moral agency, only humans could possibly have rights. In the preface to Smith's book, author Dean Koontz argues that God bestows rights, and the laws men make are simple privileges that are manipulated by those in power, "But recognizing a vertical sacred order, I must also believe that rights do not come from men or courts, or from governments, but only from God," and only God can grant or remove rights—thus animals cannot possibly be given rights (Koontz in Smith, 2012). This is similar to the Christian notion of the Great Chain of Being, a vertical hierarchy of more valuable beings, running from God first, then to angels, men, women, children, and animals. The idea of the Great Chain of Being has been quite influential, for example it shaped evolutionary science which mistakenly put, for example, some human races biologically above others (Nee, 2005).

However, we have known for some time now that social animals probably "engage in robust moral reasoning" (Casebeer, 2003) and one form of moral reasoning, empathy, has probably been passed down through evolutionary selection—and is generated in parts of the brain that are "ancient, probably as old as mammals and birds" (De Waal, 2008).

However, there are direct challenges to the HEP and anthropocentrism from **biocentric** and **ecocentric** ethics. Biocentrism believes that the value for all life, such as biodiversity, should guide our decisions. Ecocentrism argues for an even broader ethic, where the value for the holistic integrity of whole landscapes and their processes should guide our actions. However, Norton (2002) rightly points out that you can't protect biodiversity without protecting habitat and the landscape. These two approaches agree, then, on dislodging the mainstream economistic and utilitarian ethics that favor human pleasure for the consideration of other equally valuable measures. Take for example the work of United Nations Messenger of Peace, Jane Goodall. Her work, along with others like De Waal (2005), has laid waste to the reasoning behind anthropocentrism inasmuch as it insists that humans are essentially different from nature, such as our ancestors in the Great Apes. Their work implies that anthropocentrism is ethically *and* scientifically problematic, and that such an ethical frame leads to mistakes for sustainability. It should not be lost, however, that Goodall's conservation efforts have shown that local people need to have their basic needs met if conservation is to be successful (see Goodall, 2003). Valuing ecosystems or biodiversity does not mean that human interests are negated, but rather our interests exist amongst a plurality of interests of others.

Jane Goodall is a world-renowned primatologist. She has led and conducted the world's longest on-site biological observational study of more than 50 years by observing chimpanzees in Tanzania's Gombe forest, starting in 1960. As a matter of her profound experiences in the Gombe, Goodall tries to bring a message of biospheric unity to the world through illustrating the commonality between humans and chimpanzees. Chimpanzees share 99.4 percent of our genetic structure (Wildman et al., 2003), and primatologists like Goodall and Frans De Waal (2005) have argued that we have as much to learn about ourselves as we do the Great Apes when we study chimpanzees, bonobos, and gorillas, because humans are a member of the Great Ape family.

When Goodall began in the Gombe, she made an interesting and lasting "mistake"—she failed to number her subjects, as is the general practice to make the work seem "objective." Instead, Goodall *named* the chimps, and, indeed, she became closer to them. They became individuals to her—**non-human persons**. A non-human person is anything that has personhood but is not a human being. Over time, this allowed her to develop genuine relationships with the chimps that a more "objective" scientist may have closed off.

Goodall writes:

> When I first went to Africa to study chimpanzees, I had to learn to look at the world—as best as I could-through their eyes. I came to realize that we humans are not separated from the rest of the animal kingdom, that there is not an unbridgeable chasm between us and them. The chimpanzees reach out across this perceived chasm and demand that we accept them into our world or that we join them in theirs. They have taught us that we are not the only beings on the planet with personalities, minds, and above all, emotions.
>
> (Goodall, 2003)

Further,

> Once we admit that we are indeed a part of the animal kingdom, we will have a new respect for the other amazing animals with whom we share the planet. And we become increasingly shocked when we look around the planet and see what we have done to the environment. We see that our actions have destroyed the homes and the lives of countless millions of animals. And we are ashamed and shocked when we think of the way that we treat so many animals in our daily lives.

Indeed, through her research, she discredited one central pillar of the HEP. In 1960, it was believed that only humans used tools to modify the environment around us and satisfy our needs. However, early in her work, she explained how one of the chimpanzees, David Graybeard, not only used straw to fish out termites from a mound, but stripped leaves from stems to make the tool. When Goodall related this to her mentor, Louis Leakey, he famously telegraphed her back: "Now we must redefine tool, redefine man, or accept chimpanzees as humans" (Goodall, 1998).

Among other things, chimps have rich social structures, demonstrate cognition and consciousness, communicate through language and can learn new ones. They maintain personal relationships that can last 50 years or more.

They even wage deliberative attacks and war. This was a discovery that horrified Goodall. She has documented organized combat between subgroups of chimpanzees. This was not a matter of imminent danger from either chimp group, but an act of pre-meditated aggression. Incidentally, this occurred after Goodall had been with the chimps for quite some time, and she had not seen the warfare until then. Therefore, it was not a behavior that was part of daily life for the chimps, nor was it something that was part of their "nature"—it was a social decision. This is much like human war that is socially decided. In the end, Goodall tells us that chimpanzees stare at us across a void that we have created, and demand to be recognized as non-human "persons" with their own lives and purpose.

In one incident not long after she had begun to study the chimpanzees, Goodall was nearby David Graybeard and she extended to him a palm nut that he liked to eat. Graybeard took the nut and dropped it to the ground to softly hold her hand instead (Peterson, 2008). Goodall has been an ambassador for the non-human animal world, showing us we are not alone, and that if we are part of a larger family, we should act like it and be more responsible members of the biosphere.

Another very important approach is a more holistic, ecocentric, approach to values, but when we favor the biosphere, we end up favoring everything the biosphere needs in ecology—so these approaches may not be different in their policy demands.

For a holistic ecocentric ethics, usually philosophers discuss Aldo Leopold's land ethic, which argues that we should see the land and everything on it as part of our own community. Another view into ecological holism is held by many of the world's indigenous peoples, as presented in the world indigenous movement that has been building since the 1970s demanding ethical systems that are fundamentally holistic. Around the world, indigenous leaders have been organizing from many corners of the globe to articulate their dissent to the world capitalist system and the utilitarian ethics that come with it (Ridgeway and Jacques, 2013).

Indigenous theologian George Tinker argues that there are two main elements to indigenous ways of living well. In fact, the world indigenous movement, alongside a world peasant movement and world social movement in the World Social Forum, argue unapologetically with remarkable consistency that the capitalist world system and its view of the Earth and biosphere as dead matter is quite literally killing Mother Earth (see Indigenous Peoples Global Conference On Rio+20 And Mother Earth, 2012; International

Indigenous Peoples Summit on Sustainable Development, 2002; Kari-Oca Declaration, 1992). According to Tinker's schema, "Living well," or what these movements call *buen vivir*, requires the responsibility of *reciprocity*: "Knowing that every action has its unique effect has always meant that there had to be some sort of built-in compensation for human actions, some reciprocity" (Tinker, 1996). This means for everything we take from the land, we would need to pay back with something like time, and sometimes when a life is taken in Native communities while hunting, this reciprocal time commitment may mean hours or days in ceremony. Such a commitment slows the machinery of changes to the land considerably because he warns, "as far as I know there is no ceremony for clear-cutting an entire forest" (Tinker, 1996). The other aspect of *buen vivir* is living in a specific space that constitutes the community, that is, that space co-evolves with the people there and constitutes who they are as a culture and society, what practices they deem worthwhile and useful, and the values that are adaptive to that space. The community is committed to the land, plants, and animals, and everything—rocks, mountains, flowers, and eagles—are all non-human persons. People can take something from the space, indeed people must take to live, but a person or group can only take what is deemed necessary, and they must compensate the land and related non-human persons. Tinker, Vine Deloria, and many, many other indigenous leaders argue that because there is no commitment to a place and no requirement for reciprocity to non-human persons (indeed there are no non-human persons in mainstream Western ethics), Western ethics allows for wanton and unsustainable taking from the land and others who are only seen as dead matter to be used for human material welfare.

ENDPOINTS

WHAT DO WE KNOW?

We know that ethics matter to some of the most decisive elements of human sustainability, such as the distribution of food and energy. More equitable institutions, theoretically and empirically, appear to be more effective in managing natural resources.

We know that some of the most influential ethical systems, such as the HEP and the Dominant Social Paradigm (DSP), are inspired by anthropocentric priorities. In their landmark study on the DSP, Dunlap and Van Liere (1984) note that this constellation of values emerged during a time of abundance, but that it is clearly maladaptive as we move into an era where sustainability problems are more prominent. Yet, mainstream cultural attitudes continue to hold these values strongly even beyond the Western world, and groups like the Center for Human Exceptionalism openly fight directly to maintain this position.

We know that agency and moral standing matters to sustainability, because it orients our decision-making to consider more or less parts of the world. We know that if, in order to give consideration to other organisms than humans, we use a capacity test for other animals, for things like empathy, many more animals than humans would be considered. On the other hand, if we use virtue ethics, we may not need capacities to determine what we offer recognition. Remembering Chapters 1 and 2, many critical life support systems are invisible and are not openly considered in decision-making. Ethical systems that grant moral standing to a wider array of plants, animals, and landscapes may accidentally or even deliberately preserve more of these invisible critical life supports. However, ethical systems are normative, and there is no objectively right or wrong one that we can turn to in order to adjudicate human conflicts.

CRITICAL CONSIDERATIONS

Which ethical approach—utilitarian, deontological, or virtue ethics—do you prefer and why? Which approach do you think will deliver more sustainable decision-making?

What kinds of solutions to inequitable food and energy distributions do you think would work?

What rules should we make when it comes to the distribution of goods? Of hazards, like toxic waste?

Imagine you are sitting on a beach, and it is your last day on Earth. You are the last person alive, and there is nothing that you can do to improve your last moments. At that time, a sea turtle comes out of the ocean and crawls up on shore. Is there anything *wrong* with killing it? What if it were a crab? What if a mosquito were flying in a forest behind you, would it be OK to go and kill it for no reason? Would there be anything wrong with taking machinery to destroy the forest and the beach before you die? Why not? What is the moral *reasoning* behind your answer, and how does your answer impact others? Do these others matter? Certainly, these are straw-man questions, but they help you map out where you see moral standing.

WHAT DO YOU THINK OF THE FOLLOWING SUSTAINABILITY SOLUTIONS?

1. Every country should adopt a "rights of nature" approach to rules, such as the 2008 Ecuadorian Constitution, which states in Article 71:

   Article 71. Nature, or Pacha Mama, where life is reproduced and occurs, has the right to integral respect for its existence and for the maintenance and regeneration of its life cycles, structure, functions and evolutionary processes.

   All persons, communities, peoples and nations can call upon public authorities to enforce the rights of nature. To enforce and interpret these rights, the principles set forth in the Constitution shall be observed, as appropriate.

   The State shall give incentives to natural persons and legal entities and to communities to protect nature and to promote respect for all the elements comprising an ecosystem.

Since that constitutional change, its language has been, in at least one court case, upheld. In deciding a case where construction of a road was disputed, the rights to nature were brought to bear and respected. The judge's decision quoted Alberto Acosta, President of the Ecuadorian Constituent Assembly: "Man cannot survive at the margins of nature … . The human being is a part of nature, and cannot treat nature as if it were a ceremony to which he is a spectator. Whatever legal system tied to popular sentiment, sensitive to natural disasters that we, in our day, are familiar with, applying modern scientific knowledge—or the ancient knowledge of original cultures—about how the universe works, must prohibit human beings from bringing about the extinction of other species or destroying the functioning of natural ecosystems" (Daly, 2011).

2. Governments and territories should remain (or be made) small and simple, and allow stable, frugal use of nature in a way that does not disrupt dynamic homeostasis, where inputs and outputs continue in a complex system, but the system itself is stable. This means no throughput, where we simply take from nature, make something, and then leave the waste as a byproduct; and it means all political groups should remain small populations with simple rules. Without an ethic that contains these limits, humanity will suffer. Ophuls (2011) puts it this way:

> To put it more positively, ecology contains an intrinsic wisdom and an implied ethic that, by transforming man from an enemy [hell-bent only on growth, where control of nature is "ultimately a lie"] into a partner in nature, will make it possible to preserve the best of civilization's achievements for many generations to come and also attain a higher quality of civilized life.

In sum, an ethical political system would be organized and consume energy like a mature climatic system, not an immature pioneer system.

WHAT DO YOU THINK OF THE FOLLOWING SYLLOGISM?

Premise A: Institutions are required for solving sustainability problems.

Premise B: Institutions perceived unjust will be less effective for various reasons.

Conclusion: Just institutions are required for sustainability.

FURTHER READING

Dobson, Andrew. (2003). *Citizenship and the Environment*. Oxford: Oxford University Press. Dobson proposes that people should observe civic responsibilities commensurate with the harm they inflict on others, wherever the harmed live—near or very far across international borders. This harm is related to a person's Ecological Footprint, the consumption of ecological space and productivity.

Merchant, Carolyn. (1980). *The Death of Nature: Women, Ecology, and the Scientific Revolution*. San Francisco, CA: Harper & Row. Merchant examines the history of pre-Enlightenment Europe to show that the Earth was viewed as a benevolent mother figure. This organic metaphor, however, was replaced with that of a machine as thinkers like Francis Bacon saw science as a tool to force nature to reveal her secrets and control nature. This conversion of nature-as-organism to nature-as-machine was a precondition for the development of capitalism.

Vanderheiden, Steve. (2008). *Atmospheric Justice: A Political Theory of Climate Change*. Oxford/New York: Oxford University Press. Vanderheiden proposes that just institutions for climate change are essential for dealing with climate change. Justice requires that those most responsible pay the bulk of the costs and not harm either current vulnerable populations or future generations. In particular, Vanderheiden applies Rawls' "difference principle" to determine fairness, where climate rules can only be unequal if these rules benefit the least well off.

6

POLITICS AT THE END OF THE WORLD

MAP OF THE CHAPTER

Struggle and power are the storylines of this chapter, where the fault lines of sustainability from P2 provide conflicts over how much ecological protections, social equity, or economic growth are needed and through what costs. This chapter covers the politics essential to and barriers to sustainability, and discusses the nature of "development" as the central international political issue for sustainability. Included in this chapter are the dynamics of managing resources that are common to all, but can be depleted by some. Further, we cover some notions of political economy that determine what and how much is consumed from Earth's systems and cycles.

In 1895, the British Empire established the East Africa Protectorate, which became the Kenya Colony in 1920. Only about 20 percent of Kenya is arable for farming, most of this land is in the highlands, and the rest of Kenya is suitable mainly for rotating grazing. Agricultural and pastoralist tribes developed in these respective areas. British colonists appropriated the highlands for themselves, displacing primarily the Kikuyu ethnic group, setting up the conditions for the Mau Mau rebellion in the 1950s and 1960s. Kenya gained independence in 1963 and organized into a nation-state where Kikuyu ethnic leader Jomo Kenyatta and then Daniel arap Moi, both of the Kenyan African National Union (KANU) party, ruled Kenya with repressive tactics and silenced dissent for more than 40 years.

Throughout, pre-colonial, colonial, and independent Kenyan society, women did not have sufficient access to critical resources. Property was inherited primarily through the male line and women only had use-rights to land, despite the fact that they have been and continue to be the primary tillers of soil and responsible for feeding the household. In some ethnic groups, women were not even considered adults. These rules were strengthened through religion, education, and codified in the legal system, and many of these problems persist today.

To cope with these repressive conditions, Kenyan women have a tradition of forming women's work groups who share labor, for example when another woman is sick, has given birth or is dealing with other subsistence challenges. For example, when the British institutionalized forced labor alongside poll taxes, men migrated out looking for wage work, "destroying the local culture and economy and institutionalizing colonial structures and ideology" (Oduol and Kabira, 1995). From the start, the Empire's rules were meant to place Kenyan natural resources and labor into the global imperial market. For example, in 1955, the Empire made rules to put land to use for large-scale cash crops that displaced subsistence work, culture, and food sovereignty for Kenyans and resulted in "extensive overcultivation, overgrazing, and soil erosion" (Ibid). Colonial forces expected women to then add conservation to their workload, and consequently women organized open rebellion and riots, as well as other resistance efforts both during and after independence. However, some women work groups took up planting trees for their own reasons.

The National Council of Women of Kenya (NCWK) was one such group, working on improving the daily conditions for women. In 1977, Dr Wangari Maathai (1940–2011) founded what has come to be called the Green Belt Movement (GBM) and became the leader of the NCWK in 1980. The GBM responded to Kenyan women who "reported that their streams were drying up, their food supply was less secure, and they had to walk further and further to get firewood for fuel and fencing. GBM encouraged the women to work together to grow seedlings and plant trees to bind the soil, store rainwater, provide food and firewood, and receive a small monetary token for their work" (The Green Belt Movement, 2013). But, they did not stop there, they also organized and investigated why they were marginalized in their own society, fought agricultural incursions and land grabs, and worked to change the position of women *and development*, across the region (Oduol and Kabira, 1995).

Maathai used her position to confront gender discrimination from the Moi regime and the KANU party. In opposition to the Moi regime, Maathai was assaulted sometimes to the point of being beaten unconscious. She was also imprisoned, publicly ridiculed, and declared an enemy of the state (Hayanga, 2006).

Originally, the GBM was started to help provide women with what they needed most, but which had been degraded—food, water, fuel, and fodder—and the GBM knew that trees were the avenue for protecting these critical life supports. Trees keep the soil from eroding, protect watersheds, and provide important fuel for poor subsistence families, but during the colonial period and after, deforestation had been initiated in order to grow cash crops, especially coffee, for the Empire and food crops for a growing population.

Maathai recounts:

> The British government was cutting down the indigenous forest so that they could establish plantations. ... Something else I noticed. There used to be these huge fig trees. Fig trees are kind of holy trees to us ... . These trees too have been cut down, I now think that there may be a link, that these fig trees were part of a water system that kept this part of the country watered and forested. And rich, so rich. ... But now that we look back—and we are hungry, and we have no forests, and we have no water ... . We can now see what we have done in the name of development. So my childhood experience reinforces my conviction that something has gone wrong, seriously wrong in the name of development.
>
> (Maathai, 1997)

Maathai and the thousands of women who would join her, as of 2012, have planted over *50 million* trees to protect their own wellbeing and the wellbeing of their households, and later in the movement, to protect the wellbeing of an international population—all initially in the face of a repressive state, a male-dominated society, and a daunting neo-colonial environment that was being dismembered in front of them driven by the enormous power of overseas markets.

In 2002, the authoritarian Moi regime ended along with the reign of KANU, and Maathai was brought into the government and served in the parliament and as Minister of Environment and Natural Resources. Most importantly, she was awarded the Nobel Peace Prize in 2004 for her work in bringing the issues of sustainable development to the level of peace-building. Here is an excerpt of her Nobel speech available at the Nobel website:

> As we progressively understood the causes of environmental degradation, we saw the need for good governance. Indeed, the state of any country's environment is a reflection of the kind of governance in place, and without good governance there can be no peace. Many countries, which have poor governance systems, are also likely to have conflicts and poor laws protecting the environment.
>
> In 2002, the courage, resilience, patience and commitment of members of the Green Belt Movement, other civil society organizations, and the Kenyan public culminated in the peaceful transition to a democratic government and laid the foundation for a more stable society.
>
> Excellencies, friends, ladies and gentlemen,
>
> It is 30 years since we started this work. Activities that devastate the environment and societies continue unabated. Today we are faced with a challenge that calls for a shift in our thinking, so that humanity stops threatening its life-support system. We are called to assist the Earth to heal her wounds and in the process heal our own—indeed, to embrace the whole creation in all its diversity, beauty and wonder. This will happen if we see the need to revive our sense of belonging to a larger family of life, with which we have shared our evolutionary process.
>
> In the course of history, there comes a time when humanity is called to shift to a new level of consciousness, to reach a higher moral ground. A time when we have to shed our fear and give hope to each other.
>
> That time is now.
>
> (Maathai, 2004)

CHECKPOINT: WHAT POLITICS WILL MOST LIKELY AVOID NORMATIVE FAILURE?

What can we do to avoid failing the sustainability test? Unfortunately, there are no clear universally applicable ways to govern because each ecological space and political group has their own needs and histories. However, we know a few guiding principles.

First, governments and powerful actors, like corporations, cannot simply declare or promise to be sustainable, while still pursuing un-interrogated, uninterrupted, and unbalanced economic growth at the expense of dynamic, self-organizing ecological systems that take thousands of years to form. If we are not living with the boundaries of the Earth's limits now, simply rebranding policies or corporations to sound more green without actually changing the direction of consumption is not sustainable, *yet this empty promise is evident in so many government ministries and in nearly every aisle of our stores* (Smith and Farley, 2014). Corporate green washing (promising environmental sustainable practices without substance) and government policies that do not reflect the problem structure of sustainability risk the severities of Normative Failure because they do not address real problems, *and* they disarm the public to think that these problems are really being taken seriously.

Institutions at every scale, from the local to the global, will be forced to govern trade-offs and negotiate the central fault lines of sustainability. No single value of the three "Es" can dominate all the others without creating harsh imbalances that are not sustainable; and, governing systems must be adaptive. This is governance consistent with the Adaptive Cycle and Panarchy.

ADAPTIVE GOVERNANCE

Folke et al. (2005) warn, "Vulnerable terrestrial and aquatic eco-systems may easily shift into undesired states in the sense of pro-viding ecosystem services to society" and there is now wide agreement that this vulnerability, both short and long term, is a critical issue for sustainable governance. The solution to vulnerability is resilience, or the degree to which social-ecological systems can experience "recurrent natural and human perturbations and continue to regenerate without slowly degrading or even unexpectedly flipping into less desirable states" (Folke et al., 2005). **Adaptive governance** builds resilience across levels of linked social-ecological systems, and clearly this means solving very difficult collective action problems at multiple levels (see p. 154).

Scales, as noted in Chapter 2, are observed dimensions of space and time. Nearly every social-ecological system has cross-scale links and complex causes of change, or drivers. Land use change, for example, is driven by a bundle of causes across scales. Adger, Brown, and Tompkins (2005) observe: "Many, if not all systems, are inherently cross-scale and their success in promoting sustained engagement and resilient and shared management are determined by factors at a range of levels from constitutional and organizational to those at the level of resource users." A common failure in resource management and conservation comes from the failure to identify and understand the distant connections and drivers that impact the areas of interest. In part, this is also a problem of multiple actors that work at different levels and in different regimes, but that affects what happens across scales, creating constellations of actors that may have very different values, incentives, or interests. Governing multiple actors across different scales requires that the processes from one scale do not militate against the sustainable use of resources at different scales. Several processes can be at work, but two examples of problems exist where:

1 Global actors, organizations, and **institutions** can be disconnected from the local impacts of their actions, causing local problems for which global actors may not be accountable.
2 Local actions are disconnected from larger problems where local actions can add up to a global problem, but local governance often does not account for this problem. For example, local emissions of CO_2, as of this writing, have no global accountability, but together they have global impacts.

Global changes, like changes in global average temperature, affect local scales; and global institutions, like trade rules, affect the way local ecologies are managed—so land use is not only determined by local decisions, but also distant forces. This means that in a globalized world, **institutional interplay** means rules affect each other and can be reinforcing or contradictory. Thus, institutions exist in a complex human–ecological environment, where the challenge is to create **institutional fitness**, where the rules fit the ecological problem, and that are effective at each scale. At the same time, knowledge about social and ecological systems may be filled with deep uncertainties and suffer from social-environmental problems that come from converging forces. To make these abstract concepts more clear, we will look at the case of the Xingu watershed and the Belo Monte Dam.

BELO MONTE DAM AND THE XINGU WATERSHED

The Xingu Indigenous Park (PIX) in Brazil is in the Xingu watershed. Indigenous people attempted to set strong conservation strategies to preserve their forest in PIX. But, the park is surrounded by a vast agro-industrial region that is connected to the global economy through commodity chains "competing for land and water resources," and this region has caused severe erosion, seriously reduced the amount of water coming through the watershed, added industrial pollutants, and brought persistent smoke to the area that the indigenous peoples could not control (Brondizio et al., 2009). The indigenous leaders had done a good job protecting their land, but they could not control how their land was changed on the borders of the PIX.

In 2010, the Brazilian government made the situation even more tenuous for the PIX, when it approved the Belo Monte Dam, the third largest in the world, to fit into its national strategy for development. Brazil gets 80 percent of its electricity from hydroelectric dams and this is often portrayed as a "green" strategy, but the dams themselves cause irreparable damage to submerged and downstream riparian areas while displacing tens of thousands of indigenous peoples. Within a year of the approval, deforestation rates in the region doubled, and rose by a third nationally—but not solely because of the dam, rather because of the larger economic situation and laws of Brazil. These conditions are part of the institutional environment where rules need to fit the problems and scales they are meant to address if they are to be effective.

Here, the state priorities of big development projects like the Belo Monte contradict its commitments to 80 percent decline in deforestation rates by 2020 and larger stated goals to mitigate environmental problems and social inequality. Sociologists who have studied this case remark that such, "ill-conceived 'development'," is "imposed on the local people, with little effort made to consult them or even to explain properly what is happening" (Hall and Branford, 2012). The Brazilian government argues this is progress, and it will raise the local indigenous lifestyles from what are labeled backward conditions. Yet the indigenous people disagree and have argued for the last 30 years that this kind of development is devastating Mother Earth. Local Kayapó leaders go as far to say that "The world must know what is happening here, they must perceive how destroying forests and indigenous people destroys the entire world" (Survival International, n.d.).

Tropical deforestation has both immediate causes, like paying lumberjacks to cut trees, to more distant underlying causes. The latter are often not given enough consideration, but the weight of deforestation is attributed to these larger, distant economic conditions, like the way finances are traded and invested in timber or land with timber, as well the stack of rules that favor economic relationships over social or ecological relationships. Since the 1970s, political leaders have adopted institutions that move power to the economic sphere away from the state and social realms in a system called **neoliberalism**. Neoliberal policies favor removing or reducing regulations on industry and finance, and reducing state funding of education, health care, and environmental initiatives. The Belo Monte project fits within this neoliberal system. However, evidence from social science indicates that neoliberalism has caused socioeconomic crises since the 1980s: "The pattern was set for the years to come: Deregulation would lead to crisis, public authority and money would be used to resolve it, and austerity would be demanded as a way to pay for the mistakes" (Centeno and Cohen, 2012). Many scholars indicate that the forms of justice found in global environmental governance are those that are consistent with neoliberalism that favor property rights and mutually beneficial trade, and are "incapable of delivering distributive justice" and the substantive ecological limits to resource use required by the Brundtland Report (Okereke, 2007).

Still, the ecological conditions of places like the Amazon are all connected enough to affect global Earth systems and a failure to govern the Amazon sustainably affects other areas:

The Amazonian rainforest plays a crucial role in the climate system. It helps to drive atmospheric circulation in the tropics by absorbing energy and recycling about half of the rainfall that falls on it. Furthermore, the region is estimated to contain about 10 percent of the global carbon stored in ecosystems ... . *Disruptions in the volumes of moisture coming from the Amazon basin could trigger a process of desertification over vast areas of Latin America and even in North America.*

(De la Torre et al. quoted in Hall and Branford, 2012, emphasis added)

Given the work of Oran Young (2011), who has studied these environmental institutions for decades, we are presented with serious governing problems for sustainability. Young has shown that since institutions exist in a complex social and ecological setting, the challenge is to create regimes that fit the ecological dynamics that can be effective across the hierarchy of scales. Meanwhile, knowledge about social and ecological systems may be filled with deep uncertainties that interfere with actors making prudent choices.

Clearly, then, there is a conflict between what political scientists know is necessary to govern sustainably and the dominant neoliberal system, in addition to further confounding problems classic to international relations. These classic and recurring problems include a continued reticence of nation-states to commit to serious investments across scales to solve collective action problems in a world where entering and observing treaties is entirely voluntary. We will now address these **collective action problems**.

COLLECTIVE ACTION PROBLEMS

Sustainably managing common natural resources, such as fisheries, water, and forests, is essential for our long-term survival.

(Vollan and Ostrom, 2010)

Ecological systems are managed by groups of people, and some of these resources are critical life support systems, like the water cycle, and require deep coordination across scales. The more we learn about this process, the more complicated it becomes. For example, managing fisheries is much more complicated than just limiting what and how people catch fish—we also have to manage disruptions to all the processes that create the fish populations in the first place at the very same time that the political sphere is filled with local–global and public–private actors across space and time with different goals and influence. First, we must explain the standard lessons of the "**tragedy of the commons**."

Common land of open pasture systems, "probably antedate the idea of private property in land, and are therefore of vast antiquity" (Taylor quoted in Buck, 1985). Peasants and indigenous peoples have held land in common before the idea of private property was ever conceived. The French liberal philosopher Rousseau imagines the critical moment when this changed:

> The first man who, having enclosed a piece of ground, bethought himself of saying This is mine, and found people simple enough to believe him, that man was the real founder of civil society. For how many crimes, wars, and murders, from how many horrors and misfortunes might not any one have saved mankind, by pulling up the stakes, or filling up the ditch, and crying to his fellows, "Beware of listening to this impostor; you are undone if you once forget that the fruits of the earth belong to us all, and the earth to no one!"
>
> (Rousseau in 1775)

Given that all of Earth's ecosystems are now dominated by human actions (Vitousek et al., 1997), the commons of the Earth—the World Ocean, the atmosphere, the web of life and its gene pool, terrestrial ecosystems like forests and grasslands, groundwater—are being consumed with alarming thirst. At least some of these commons, such as the sea floor outside national management areas of the World Ocean, have been referred to as the "common heritage of mankind" that should be used and protected for the long-term interests for all people. One tragedy is simply the loss of abundant commons that allowed collective groups of people to subsist with deference to the community.

Soron and Laxer (2006) write, "Generally, the commons refers to those areas of social and natural life that are under communal stewardship, comprising collective resources and rights for all, by virtue of citizenship, irrespective of capacity to pay," but once commons are enclosed for private use it is the capacity to pay and be paid that concentrates power with those who have claimed commons for their own. What remains commons must be governed by collectives.

At least since the time of Aristotle, social science has wrestled with collective action problems. Generally, these are the problems that arise when individuals resist cooperating with each other in order to provide a common good for everyone.

And, it gets worse. Individuals may even work to *destroy* critical resources everyone needs for their own short-term gain—this is what Garret Hardin (1968) named **tragedy of the commons (TOC)**. Hardin's article in the journal *Science* may be the most cited article of contemporary science, even though it contains several important errors. Hardin invites us to imagine an English pastoral commons. On this commons, there are no rules. He assumes that all people are strictly rational, so a herder will realize that if he puts an extra sheep on the pasture, he will realize the benefit of one full sheep for consumption, trade, etc. The cost, however, will be shared by the whole community dependent on the pasture—so the costs of overuse are spread around, and the benefits are concentrated. Because Hardin believes everyone is rational, everyone will come to the same realization and they, too, will add another sheep.

When it is obvious that the pasture is doomed, the herders do the unthinkable. They *intensify* their use because they know that the pasture will be used up, and each one wants to beat their neighbor to using the last bits. This resource race to the bottom creates a lock because everyone now is in this race to use the resource before it is gone.

When the field is over-grazed, no one can use the resource as it has collapsed—and this is the tragedy.

We can argue against Hardin's assumption of rationality since humans are not rational in the way he proposed. But, this does not matter to Hardin's case, because all we need are *enough* exploitive rational utility maximizers and they will use up the resource even if there are altruists who opt-out of the race. The utility-maximizers will take the altruist's place. This is why Hardin notes that the conscientious will suffer the same fate as everyone else, because even if some opt-out, the commons will still collapse under this situation. Related, one of Hardin's central concerns in the TOC paper was population growth in a limited world. If we appealed to people's sense of conscience to have less children, those who did not respond—the less conscientious—would then eventually become the genetic stock, and the conscientious would have eliminated themselves.

In the end, Hardin, argues that all commons are destined for collapse and that, "Ruin is the destination toward which all men rush, each pursuing his own best interest in a society that believes in the freedom of the commons. Freedom in a commons brings ruin to all" (Hardin, 1968). Hardin then suggested that in order to avoid this fate, the commons should be privatized to the greatest extent possible, and what cannot be privatized should be managed by an authoritarian state ("mutual coercion, mutually agreed upon"). William Ophuls followed Hardin by proposing that we are at a critical juncture for global sustainability because the world economy was promoting constant growth and more freedom to consume more, and this was going to lead to dreadful scarcities of critical goods, like food. Thus, the freedom to destroy our commons, he argued, will lead to stark despotic and repressive political life. While a more permissive set of freedoms may have worked in a period of unparalleled abundance, Ophuls noted we are entering the "scarcity society" where the ecological crisis imposes drastic limits on our freedoms. In this case, removing some freedom now, would preserve some freedom later (Ophuls, 1974).

HARDIN'S ERRORS, OUR HOPE

Hardin made at least two important mistakes that help us understand our situation a little better. The first is that he confused resources for the rules of using resources. The second is that the historical conditions of the actual English countryside were very different than he supposes, and these differences provide insight into how people manage real common pool resources.

To understand Hardin's first error we turn to the late Nobel laureate Elinor Ostrom (1933–2012), who studied actual common pool resource use around the world and discovered that people do not always destroy these resources. One mistake Hardin made was to confuse resource systems (the lakes, oceans, atmosphere, pasture, etc.) for the rules of those systems (**regimes**, or institutions). Hardin was really explaining an open pool regime, which is an institutional setting where there are no rules of access or use to limit consumption of the pasture, but open systems are not the only option for common resources. Common pool resource regimes, however, tend to have rules, and there are a number of different kinds of rules that people have used to allocate shared resources, especially property rights and responsibilities.

There are several different kinds of property rights (Ostrom et al., 1999):

1 Open access: where no property rights are enforced.
2 Private property: individuals hold rights to the resource and may exclude others.
3 Group property: where a group of people may form collective use rules for the good, and disposal of the property requires group agreement.
4 Government/state property: where the resource is regulated or subsidized through the state, like a national forest or a national park.

These property rights make up part of the regime that applies to a resource. There are four types of goods/resources, determined by how exhaustible the resource when it is consumed and how hard it is to exclude other users (see Table 6.1). Table 6.1 delineates pure types, but many resources may share qualities of different types, or the nature of its exhaustion or exclusion may have gradations.

Thus, private property is used by individual owners and is immediately excludable and exhaustible. A public good is one that markets have a very hard time providing because there is no scarcity that comes from consumption of the good, and it is hard to keep people from using it without paying. For example, when I use national security, it is not depleted; and, my nation will provide national security whether I pay my taxes or not.

Common pool resources are the most important for sustainability because these include our environmental resources that are hard to regulate but that are depleted when they are not regulated. This illustrates the tension between sustainability and the neoliberal economic system that favors deregulation.

The atmosphere, as of this writing, is an open regime for a common pool resource, and countries are using this sink for carbon dioxide and other greenhouse gasses (GHGs) just like the herders

Table 6.1 Basic types of resources

	Excludable	*Non-excludable*
Exhaustible/rivalrous	Private goods	Common pool goods
Non-exhaustible/non-rivalrous	Club/toll goods	Public goods

used up the pasture in the TOC, with some exceptions where emissions are traded regionally as in the European Union. Climate change, then, is a classic tragedy of the commons, and research in political science tells us that this resource needs more appropriate institutions. At the same time, nation-states have sovereignty and can decide to cooperate with other countries, or not; currently countries are cooperating to do almost nothing to mitigate GHGs.

Hardin was applying open access regime to what was really a common pool regime. Open access is patently unsustainable, making the TOC a perennial authentic concern and where open access of common resource systems exists they will undoubtedly be exhausted. However, if there are group property rights in Hardin's pasture, the group can impose restraint on more herders coming into the pasture and from one herder **free riding** on the rest of the group. A free rider is an individual who benefits from a publicly provided resource without paying. Famed political economist Mancur Olson (1932–98) noted that it is difficult for a large group of people to solve their collective action problems because of free riding since free riding is preferable to paying. Indeed, in the problem of climate change, many nation-states prefer to be free riders to agreements limiting GHGs placing the cost of cooperation and a stable climate on others; the fact that these nation-states cannot be compelled by any super-national entity makes the problem of climate change one of the most serious collective action problems we must solve for long-term sustainability. Notice, however, the world economy is governed by a super-national authority in the World Trade Organization, which sanctions defectors even if those countries are not members, and this kind of authoritative governance has been achieved and is therefore possible.

Thankfully, "a basic finding [of contemporary social science] is that humans do not universally maximize short-term self-benefits, and can cooperate to produce shared, long-term benefits" (Vollan and Ostrom, 2010). So institutions are required to solve collective action problems, and these rules form an important part of **social capital**, or, "value of trust generated by social networks to facilitate individual and group cooperation on shared interests and the organization of social institutions at different scales" (Brondizio et al., 2009). Social capital allows for problem solving to occur, and builds up trust in resource communities that work against the problems of short-term maximizing. It is possible that through cooperation, knowledge in the respective community could change in ways that influence more conservative protections. But, if social capital is weak, opportunities for short-term maximizing grow. In addition, social capital may work regressively. "Here social capital is more than social networking: it is the establishment of social relations that create systems of power that are exclusionary, self-supporting, and based on uneven social conditions" (Gareau, 2013). There is no reason to believe that networks of knowledge will work for the benefit of everyone, where, for example, US hate group the Ku Klux Klan probably mobilized a lot of social capital within its ranks, thus the kind of social capital, who benefits from it, and how it is organized are important considerations.

This brings us to another important error of Hardin's TOC—historical fact. The English pastoral commons in medieval and post-medieval England were, *in fact*, sustainable for hundreds of years (Buck, 1985). The reason they were sustainable was because there were institutions that kept Hardin's scenario from becoming reality, including limits to heads of livestock that were, in part, instituted from an understanding of **carrying capacity** by the peasants who managed these lands. Later, these commons were sent into serious decline, but not from a TOC. The English commons were sent into decline because of the Enclosure Reforms that fenced off the commons by English elites. The peasantry were dispossessed of their common property and means of subsistence, and the change in agricultural technology allowed for these commons to be used more intensely by the landed elite who gained the land (Buck, 1985).

In summary, several tragedies have unfolded that put communities in the modern era at risk. The first tragedy is the loss of commons that are the subsistence base for all people, and that this loss is one that generally puts vulnerable people in more jeopardy while empowering elites who have little incentive to keep the ecology intact, especially since extracting resources from these areas will increase wealth and power and allow these individuals, countries, or firms to move on to the next fishery, forest, or fen (marshes).

The second tragedy is that the human family lives in multiple layers of collectives, such as different countries. These different collectives may find it in their short-term interest to destroy critical life support systems, and that in order to preserve the commons for our long-term survival, cooperation and restraint are mandatory. However, two layers of our collectives militate against cooperation and bristle at the normative requirement of restraint: a neoliberal world capitalist economy, and the system of nation-states built on a concept of **sovereignty**, or non-interference with rights to control state populations and resources. Firms in the capitalist system work to accumulate capital and therefore consume ecology, and their interest is in profit, revenues, and growth, not restraint or stasis. Countries often imagine themselves in an anarchic situation where any cooperation they commit to is met with exploitation by free riders and a loss of their sovereignty, and this has created serious obstacles to global environmental governance.

THE TRAJECTORY OF GLOBAL GOVERNANCE FOR SUSTAINABILITY

The story of Dr Maathai, the first woman in East Africa to earn a Ph.D., a woman who organized other women to fight desertification with tens of thousands of tree-plantings in a male-dominated society, who then goes on to win the Nobel Peace Prize, tells us that one person can make a difference. It tells us that *social* movements *can* matter (see p. 168), because if the effort simply remained only the individual work of Dr Maathai and not the broader GBM, the impact would have been so much smaller. It tells us that in development politics, gender matters but is often ignored as a fundamental social force. It also tells us that, because her story is so remarkable, that such stories are not "normal" they are exceptional. Further, the story and related scholarship in the area tell us that some of the most important politics for sustainability are those about development, illustrated by the Belo Monte Dam case.

In this section we will trace the effectiveness of some of the more important tools for global environmental governance, starting with the Brundtland Report that was published in 1987. This report inspired the United Nations to organize the UN Conference on Environment and Development (UNCED), or the Rio "Earth" Summit in 1992.

Scholars appear to agree that whatever commitments the international communities had in 1992 have dissipated, and, in fact, the United Nations itself determined that the urgency of 1992 had faded within five years. The global enthusiasm related to UNCED produced the Convention on Biological Diversity (CBD), the United Nations Framework Convention on Climate Change (UNFCC), Agenda 21, the Declaration on Environment and Development/"Earth Charter," and the Non-legally Binding Authoritative Statement of Principles for a Global Consensus on the Management, Conservation and Sustainable Development of All Types of Forests. We will discuss each briefly in turn.

The CBD is one of the most ambitious and important environmental regimes. It calls for countries to make national plans that include protected areas of land and sea, restoration of habitat, and scientific monitoring and research to reduce the loss of biodiversity. However, it was clear by 2010, that the CBD had failed to reach its targets, so in a meeting at Nagoya, Japan, new targets were put in place to conserve 17 percent of terrestrial and inland water areas, 10 percent of marine and coastal areas, and restore 15 percent of degraded areas. Biologists, however, note that these goals fall "woefully below" what is "ultimately needed to sustain life on Earth" (Noss et al., 2012). Worse, the new commitments *between* countries will only exist on paper until strong conservation institutions exist *within* the countries and are observed by most actors. Prior research indicates that strong institutions are needed from the local community level up through the national and international scales where financing, incentives, and knowledge favor conservation, but these cross-scale institutions are found wanting (Barrett, Brandon, Gibson, and Gjertsen, 2001). In some cases, the CBD rules conflict with more powerful institutions like the World Trade Organization; and, some question the use of neoliberal values and market-based tools favored by the CBD as feeding into the larger cause of the problem (Corson and MacDonald, 2012). The 2010 goals are too new to really know if they will work, but without a comprehensive and integrated set of strong local–national–global decision-making procedures for land use, invasive species, climate change mitigation, hunting, and pollution, biologists warn we face a "tsunami of extinction" that washes away everything from predators like lions and tigers, to entire biomes like coral reefs, grasslands, and specific forests (Lovejoy, 2012). Some scholars find the loss of biodiversity so daunting, they have speculated on potential cascading events that eliminate the human species in what would be the "seventh mass extinction" (Carpenter and Bishop, 2009); thus, while we don't know what will happen in the future, it is safe to say that the CBD has not yet interrupted the death march of the Sixth Great Extinction.

Regarding climate change, the UNFCC's most powerful tool, the Kyoto Protocol that put limits on emissions from participating wealthy industrialized countries, expired in 2012, and there is no serious institution to fill in the gap. It is likely that the global accord most impacting greenhouse gases now is the Montreal Protocol on Substances that Deplete the Ozone Layer because some ozone depleting substances are also greenhouse gases (Velders et al., 2007). The Kyoto Protocol failed because it could not overcome the commitment to growth in energy use in just about every country, the priority for cutting emissions could not compete with economic priorities, where in the US and UK anti-environmental programs became popular, denying the authenticity of climate science, and in rising economies like Brazil and China, reducing emissions could not compete with national development goals.

Now let us discuss Agenda 21. The organizing committees to the Earth Summit began developing a plan of action for sustainability, named Agenda 21, which was approved by most countries of the world in Rio, even though it was "shaped largely by Northern elites" (Doyle, 1998). Agenda 21 advised ways to interact with wetlands, forest, coasts, and other areas to promote sustainable development and there are now perhaps hundreds of cities that have put in place Agenda 21 plans. Agenda 21 is a complex, expensive, but voluntary set of suggestions, which various localities can choose to adopt at varied levels of commitment. However, it is clear that Agenda 21 does not deal with the actual problem structure of sustainability: both consumption patterns and population were left out of the massive 500-page document. In particular, Agenda 21 removed references to contraception at the behest of the Vatican and the Philippines, while also neglecting militarism and international debt because they were simply too controversial (Dresner, 2008). Worse, some analysts believe environmental advocates should reject Agenda 21 out of hand for its lack of potential for real advances toward sustainability because it is seen as a co-optive tool of the neoliberal system:

> Agenda 21 has also been successful in selling a concept of sustainable development which continues to promote the Enlightenment goals of progress through economic growth and industrialisation at all costs. But it is worse than this: it also advances the globalisation of radical liber-tarian market systems, along with US style "apolitical" pluralist systems of democracy.
>
> (Doyle, 1998)

Thus, critics like Doyle indicate that Agenda 21 has not only failed to politically deal with the causes of unsustainability, but Agenda 21 actually participates in reproducing causes of inequality and global environmental change.

When it comes to sustainable forestry, the UNCED agreement is hardly worth mentioning, and governance in this area has hit a noticeable wall. For example, from the 1990s, many international forestry governance projects, like the World Commission on Forests and Sustainable Development, which have come and gone because they have lacked any serious political support to really limit timber harvests or interfere with deforestation. Now, most international forestry governance simply hinges on weak and ineffective voluntary measures. David Humphreys (2006) writes in *Logjam: Deforestation and the Crisis of Global Governance*:

> This is so that such measures will not be ruled as barriers to trade under World Trade Organization (WTO) rules. The WTO has established a neoliberal [an economic system that places free enterprise and trade above all other considerations] under which free international trade and other neoliberal objectives trump public goods provision. Trade restrictions to ban the international trade of unsustainably managed or illegally

harvested timber are inadmissible ... . So, ... deforestation, especially in the tropics, continues largely unchecked.

Humphreys warns that unchecked deforestation means that by 2050, 40 percent of the Amazon rainforest, one of the most rich areas of biodiversity in the world, will be lost along with 25 percent of the mammals living there. There is approximately 32,688,000 km^2 of forest around the world, and using modern satellite techniques, we know about 1,011,000 km^2 were lost only in the 5 years between 2000–2005 (Lovejoy, 2012). However, governance needs to manage something more complicated than "simple" deforestation, because there is now evidence of a global die-off of forests from climate-related changes to hydrologic cycles (Garcia and Rosenberg, 2010). Human-induced climate change is affecting the timing, length, and strength of droughts, and a 2011 *New York Times* article reported the following:

> The devastation extends worldwide. The great euphorbia trees of southern Africa are succumbing to heat and water stress. So are the Atlas cedars of northern Algeria. Fires fed by hot, dry weather are killing enormous stretches of Siberian forest. Eucalyptus trees are succumbing on a large scale to a heat blast in Australia, and the Amazon recently suffered two "once a century" droughts just five years apart, killing many large trees.
>
> (Gillis, 2011)

Later studies have shown that most trees require a delicate balance of water provision, dependent on stable climate conditions, and this may explain part of these die-offs which make trees more susceptible to bark beetle infestations, disease, and fire (see Anderegg et al., 2012).

Together, the Rio projects provided an opportunity to build problem-solving networks across countries, and experiment with governing global environmental change problems. However, at this point, scholars tend to agree that "Rio environmentalism" is "dead" and the energy for these global governance strategies evaporated in the following decades (Park, Conca, and Finger, 2008). DeSombre (2006) notes that the declarations from UNCED "reflect the dominance of the liberal economic order," where economic growth is the dominant goal, and viewed as compatible with environmental problem-solving, and there is little evidence of a coherent set of global institutions working toward restraint, justice, or foresight.

That said, there are more optimist scholars who comment that UNCED provided important lessons from its failures, provided social capital and networks of people who can now act more effectively against global environmental and development problems— but only ten years after UNCED at the World Summit for Sustainable Development, "a serious environmental agenda was almost entirely missing," where leaders could not even affirm the goals of the original Rio meeting, let alone drive new innovative initiatives. The situation appeared even worse at the Rio + 20 meeting where the outcomes of this meeting hardly met journalistic standards for a news story.

Park et al. (2008) ask,

> How is it possible—15 years after UNCED, 20 years after the Brundtland Commission, and more than 30 years after the Stockholm UN Conference on the Human Environment—that the great global challenge of securing the ecological future of the planet and its people has reached a point of such political and social insignificance?

They believe UNCED and the cautious optimism of the 1992 Rio meeting failed for three important reasons that are reflected in consistent findings in global environmental political science:

1 A total underestimation and failure to deal with a globalized industrial economy.
2 A misunderstanding of environment–development problems.
3 Deep social conflicts and the problems of authority or ability of states were ignored.

In other words, even though UNCED was the *most ambitious* global environmental moment, it did not deal at all with the "globalization of production systems, the rising power of the transnational corporate interests, the diffusion of consumerist tastes and lifestyles, and the weakening of national regulatory systems" that have been growing at least since the 1970s within the neoliberal world capitalist system. Also, UNCED focused mainly on country agreements about transboundary problems, but much of the action needed to be a coordinated effort to reduce impacts at the local levels. For example, in thinking only about transboundary problems (such as river cooperation), it failed to deal entirely with the changes to freshwater *systems* that are largely controlled domestically, inside country borders apart from any international negotiations or diplomacy. Finally, UNCED overestimated the ability and willingness of countries to remedy issues for global sustainability, with a number of declarations that start with "states shall … despite precious little evidence of their effectiveness and despite the fictional world of sovereign diplomacy presumed by such efforts" (Park et al., 2008).

There appear to be few sober voices to tell us we are making good international progress to ensure a safe living space for humanity. There are serious conflicts between scales, actors, and institutions, interplay between rules can be contradictory, organizations related to solving these problems have been limited, and the larger political economic demands appear to dominate policymaking environments. Tapio Kanninen (2013) assesses the situation and concludes, "neither the present UN system, the G20, nor other existing intergovernmental institutions have developed systematic and credible mechanisms or provisions to respond effectively to a global emergency," and in turn believes that the entire constitutional structure of the UN needs to be remade in order to make a Great Awakening [emergence of a global endorsement for necessary changes toward sustainability] possible in order to avoid the Great Disruption [civilization collapse/systemic social crises]. Even more optimistic voices put the issue plainly:

> Without apologies for dramatic effect, the main issue we are concerned with is nothing less than the survival of a biosphere safe for human life. Our life support systems are under dire threat, and the interdependent character of nature will not permit any easy escape. This problem reflects a crisis in governance as much as it does biological and climactic unbalance, and it is arguably, the task of global ecopolitics to fix it.
>
> (Stoett, 2012)

Stoett believes "we can do this" but the international community will need to improve global governance by paying more attention to the issues of justice, which garners more ardent participation and to work on multi-actor, adaptive governance. Consideration of the least well off and most affected by global environmental changes is limited, and the distribution of wealth and opportunities appears to be passing regressively from the most vulnerable to the most well off, especially in land use change and climate change (Vanderheiden, 2008).

However, what would inspire governments to take on large risks and expense to initiate large scale changes to energy systems, food systems, and the built environment in collaboration with other governments? This raises the issue of social movements and civil society that started this chapter.

PEOPLE GET READY: CIVIL SOCIETY, SOCIAL MOVEMENTS, AND HEGEMONY

When we started this chapter, we learned about the GBM that has had an enormous impact on sustaining the lives of women in Kenya by improving the ecological, social, and economic conditions in which Kenyan women live. The GBM was and is a part of **civil society**, that part of society that is neither in charge of state governments nor large corporations and economic production. These three major social groups struggle in **arenas of domination**, that political space where state, corporate, and civil society projects vie for influence not just over the nuts-and-bolts of policy, but over the broad moral landscape in which policy is made (Migdal, 1997).

When it comes to sustainability problems like mitigating climate change, some scholars like Stephen Hale (2010) argue that nation-states have been ambivalent, and corporations have been outright hostile to making necessary changes; thus, it is up to the "third sector" of civil society. Civil society is formed by informal and formal groups of people meeting in public for recreation, helping others, solving problems of common interest, and making demands on governments and corporations. When there is a "a conscious, collective, organized attempt" to make demands on society, we call this a **social movement** (Wilson, 1973). Social movements typically organize to make demands for change, and therefore are usually formed at the grassroots; and, if these efforts to demand change threaten to be successful, those who benefit from the status quo arrangement, elites, may organize an opposing, or counter-movement, to stop or reverse this change. Social movements are not always progressive, and countermovements are not always reactionary, but the conditions lend themselves to these patterns.

Social movements have been successful in the past, such as the Environmental Movement in the US during the 1960s and 1970s that resulted in many laws that established environmental protections and standards for air and water quality.

However, we must guard against the notion that civil society immediately and always is successful or progressive. Several problems confound this simplistic expectation, including the **structure of political opportunity**, limited **autonomy** and **popular sovereignty**, and **hegemony**.

The structure of political opportunity is made up of contextual conditions: the capacity and organization of nation-states, critical events that focus public attention and concern, and the arrangement of interests of different groups, including the availability of alliances. Groups in civil society can change or take advantage of opportunities to be more effective, but just because civil society groups have a grievance or want something, does not mean there is a political opportunity for those demands to be realized.

The potential for civil society to advance change are also bound by its autonomy and popular sovereignty. When observing civil society in the US in the early 1830s, Alexis de Tocqueville observed, "The people reign in the American political world like God over the universe. It is the cause and aim of all things, everything comes from them and everything is absorbed in them." Today, in Western democracies there is sometimes the assumption that the people rule as effectively as de Tocqueville estimated here, and this influence is referred to as strong popular sovereignty. Further, he believed that the people who made up informal and formal affiliations in public were authentically organized by and for the people's interest, which in the majority would be concerned about the overall welfare of the nation. In other words, de Tocqueville did not think that civil society was organized to defend ruling class interests but the interests of the people, and to the extent that civil society is organized around its own interests, it is autonomous from the other forces in the arenas of domination—it is not a puppet of the state or corporations.

Can civil society, then, organize to demand nation-states and corporations shift their mode of operating in ways that build a better future for everyone? Few political analysts would say that civil society is never sovereign or autonomous, but some think that the potential for progressive organizing against maladaptive governing systems faces steep challenges. While de Tocqueville saw the people as "God" in governing in nineteenth-century America, in twentieth-century Italy, Antonio Gramsci (1920s) provides us with a more skeptical view.

Gramsci (1996), a Marxist, believed that the ruling class governed society through hegemony. Hegemonic power is power that the people do not generally question, but which embodies rules of society that are normalized to actually reinforce the interests of the ruling class through the state. Hegemony is formed by the combination of coercive power of the state plus the normative power of civil society (churches, schools, parties), and thus Gramsci believed civil society was quite often another arm of the state. According to this model, civil society has little autonomy or sovereignty. When the state wants to enact change, it organizes civil society to generate support for its programs and "educates" the educators who spread a message that is ultimately in the interest of ruling class elites.

In order for civil society to make a difference and challenge social values in the arenas of domination, civil society must be at least partially autonomous and sovereign. In particular grassroots, transnational movements like GBM will need to move the broad moral landscape, especially when confronting the neoliberal system, which has generated countermovements against global environmentalism through the Environmental Skepticism and Climate Denial Countermovements. These countermovements challenge the authenticity (reality and importance) of global environmental changes, but are organized by ideological think tanks that work to defend against ecological and justice demands against regulations and global capitalism. Environmental Skepticism as well as Climate Denial appear as hegemonic forces described by Gramsci, and to be effective, grassroots environmental movements will need to counter this hegemony across national borders to challenge the nature of consumption, and the current modes of economic power and production, and strategically use its consent.

ENDPOINTS

WHAT DO WE KNOW?

In 2012, Earth System Governance Project, which is the largest social science research network dealing with environmental governance—some 1700 collaborators from around the world—speaking from a research program that took ten years mandated from the International Geosphere–Biosphere Program, the World Climate Research Program, Diversitas (a worldwide program to study biodiversity), and the International Human Dimensions Program on Global Environmental Change issued this statement prior to the Rio + 20 Meeting, in hopes that members to that meeting would initiate serious changes:

> Our research indicates that the current institutional framework for sustainable development is deeply inadequate to bring about the swift transformative progress that is needed. In our view, incrementalism—the main approach since the 1972 Stockholm Conference—will not suffice to bring about societal change at the level and speed needed to mitigate and adapt to earth system transformation. Instead, we argue that transformative structural change in global governance is needed.
>
> (Biermann et al., 2012b)

We know now that the 2012 Rio meeting did not even come close to these urgent goals. Political scientists and related scholars researching global environmental governance agree that the current institutions governing global environmental changes, such as biodiversity loss or climate change are "deeply inadequate," and that "transformative structural change" must occur to address rapidly growing global crises. This report is very clear: we are not governing toward sustainability.

CRITICAL CONSIDERATIONS

What should the various roles of governments be at different levels? What is the role of government when it comes to putting rules in place for economic activity? How would these rules work with international firms or international financial organizations like the WTO?

What do you think the role of inequality plays in adaptive governance? What kinds of rules do you think work better at a local level, and what rules do you think work better at international levels?

How does science and knowledge interact with politics and building workable institutions?

WHAT DO YOU THINK OF THESE SUSTAINABILITY SOLUTIONS?

For every claim of "sustainability," governments, corporations, and civil society groups should be obligated to show how their policies, products, or programs reduce global consumption, increase social equity, and improve material wellbeing. Violators could be subject to public shaming campaigns or publicly asked to reform their work to make them really sustainable.

Earth System Governance Project offers the following nine-point plan from two reports (Biermann et al., 2012a; Biermann et al., 2012b), below:

1 "*Strengthen international environmental treaties*" (2012b) so they are more effective: have precise ways to know if they are working, that the rules fit the problem, processes are flexible, can incorporate new scientific lessons, among other more detailed strategies.

2 "*Manage conflicts among international treaties*" (Ibid), where, "integration of sustainability policies requires that governments place a stronger emphasis on planetary concerns in economic governance"(2012a).

3 "*Fill regulatory gaps by negotiating new international agreements*" (2012b).

4 "*Upgrade UNEP (United Nations Environment Programme)*" to a "specialized UN agency for environmental protection, along the lines of the World Health Organization or the International Labour Organization" (2012b).

5 "*Strengthen national governance*" through new policy instruments alongside the existing regulatory structures (voluntary agreements have not delivered their promises without being embedded in regulatory structures).

6 "*Streamline and strengthen governance beyond the nation state*" to improve the transparency and reporting of efforts that are beyond the reach of national governments.

7 *"Strengthen accountability and legitimacy"* (2012b): where, for example, "stronger consultative rights for civil society representatives in intergovernmental institutions would be a major step forward, including in the UN Sustainable Development Council that we propose" (2012a).

8 *"Address equity concerns within and among countries"*: "equity and fairness must be at the heart of a durable international framework for sustainable development. Strong financial support of poorer countries remains essential" (Ibid).

9 *"Prepare global governance for a warmer world"* (2012b) because completely stopping global warming is not unattainable; thus, adaption in cities and areas where vulnerable populations reside, as well in overall energy, food, and water systems, need to be prepared for unexpected shocks.

WHAT DO YOU THINK OF THE FOLLOWING SYLLOGISM?

Premise A: The nation-state is the highest level of coercive authority.

Premise B: Nation-states thus far have systematically failed to initiate effective adaptive governance for sustainability and appear ambivalent to the challenge.

Conclusion: Nation-states have lived out their useful (adaptive) function for humanity and a new kind of organization is required for the continuity of social-ecological systems.

FURTHER READING

National Research Council. (2002). *The Drama of the Commons. Committee on the Human Dimensions of Global Change.* Edited by T. Dietz, E. Ostrom, N. Dolsak, P. C. Stern, S. Stovich, and E. U. Weber. Division of Behavioral and Social Sciences and Education. Washington, DC: National Academy Press. This compilation provides an authoritative review of what we know about common pool resource management and solving collective action problems of sustainability.

Scott, James C. (1998). *Seeing Like a State: How Certain Schemes to Improve the Human Condition Have Failed.* New Haven, CT: Yale University Press. This book details centralized development projects that fail, causing misery and death, because they impose demands that ignore the complexities and interdependencies that exist on the ground locally.

Young, Oran R. (2013). *On Environmental Governance: Sustainability, Efficiency, and Equity*. Boulder, CO: Paradigm Publishers. This book provides a lucid and easily accessible review of the best research on global governance challenges and solutions.

7

THE COLLAPSE OF CIVILIZATIONS AND DARK AGES

MAP OF THE CHAPTER

The final storyline of the book is the role of hubris, because the most obvious pattern of collapse and Dark Ages is that they recur across time, yet we see little evidence of applying these lessons to contemporary social decisions. Across the ages, the mightiest civilizations have collapsed, often because of political failure. Deeply complex societies, such as the Hohokam, have come and gone, though the causes of theirs and others are subject to a great deal of debate. This chapter will first report on archeologists who warn against over-simplifying collapses. Then the chapter will explain plausible theories of collapse and then cover the Lowland Maya collapse as a specific case. We then turn to the perhaps more important phase of collapse, Dark Ages, which are systematic social crises across world-systems that last at least 500 years.

Sometime 4000 years ago in the Sonoran Desert of the present-day southwest United States and northern Mexico, someone planted what probably was the first maize (corn) in North America; this is sometime after the earliest domesticated maize was planted south of this region somewhere in Mexico between 6250–59,000 years before the current era (BCE) (Jaenicke-Despres et al., 2003). Later, the people of this Sonoran region would build elaborate irrigation canals that spanned more than 73,000 km².

The Hohokam knew enough about the cycles and dynamics of the Sonoran Desert to sustain year-round agriculture. The Hohokam had learned how to effectively separate crops to keep bees from cross-pollinating unintended flowers, they knew to rotate crops to avoid soil exhaustion, and they knew—critically—that their irrigated fields had to be drained to avoid them from salinizing.

Anthropologists Suzanne and Paul Fish comment on the remarkable resilience of Hohokam society:

> The Hohokam are especially notable for the long-term continuity of their lifeways … . Once established, some clusters of dwellings in the largest settlements persisted—renovated, extended, and rebuilt—up to several hundred years. Central plazas in these foremost settlements remained the heart of village from beginning to end … . Settlement stability was an outcome of the productivity and sustainability of Hohokam agriculture.
>
> (Fish and Fish, 2008)

Indeed, avoiding salinization is a notable feat in itself. Salinization of soil is one factor early civilizations, such as Mesopotamia, had to deal with in the face of rising populations and fluctuating climatic trends that were, in conjunction with administrative/political failure, responsible for civilization collapse in the Near East.

The Hohokam built ball courts, "Great Houses" (large intricate buildings), and platform mounds like pyramids that all indicate a complex society had grown around the rivers of the Sonoran Desert, and then later had spread out to upland areas of present day Arizona, e.g. Flagstaff. Yet, the Hohokam did this all without a steep hierarchy. No gravesites have ever been found that indicate steep inequality, authoritarian rulers, or even a bureaucracy. They did not develop a state that, in other arid societies of the world, evolved around control of irrigation as a way to control the population (Worster, 1985).

Further, the "Hohokam" label is probably more of a convenient term to refer to what probably were multiple social groups. When the Spanish arrived, there were several languages and lifeways that indicate the Hohokam may have been a heterogeneous set of ethnic groups—yet there was enough unity through the monument centers that communal life was coordinated. It is possible that the Hohokam were made up of different ethnic groups that cooperated enough to reproduce similar workable and adaptable patterns across the region, such as sharing crops.

And, yet, the Hohokam do not exist in this way anymore.

Anthropologist James Bayman (2001) divides the span of Hohokam history into four sections: the Formative, Pre-Classic, Classic, and Post-Classic periods. The formative early agricultural period of the Hohokam lasted almost 2000 years from 1000 BCE–700 CE, which was able to endure as a flexible and stable set of societies spanning at least from what we now know as greater Phoenix and Tucson, Arizona, USA. However, the earliest sites of the Hohokam maize agriculture date back to 1700–1200 BCE, which means that the contemporary O'odham people and other potential tribal ancestors to the Hohokam still in the southwest have an agricultural history that dates back almost 4000 years.

During the Pre-Classic Period of 700–1150 CE, the Hohokam regional groups developed more centralized communities that were spread out farther from the original settlements along the Salt, Gila, and Santa Cruz rivers. At this time, population increased.

The Classic Period lasted from 1150–1450 CE, where the network of canals was extended to the Tonto Basin and cultivation and consumption increased during this period, but health began to decline in some areas and signs of nutritional stress were evident as people began to eat smaller and smaller fish. Meanwhile earthen monuments, called platform mounds, were built around the region that some scholars believe were used as residences and ceremonial centers. These centers conveyed hereditary leadership, but were also centers for mediating different ideological and cultural elements of Hohokam communities. Evidence indicates that villages engaged in craft-based specialization and trade, but this trade did not come with high levels of social inequality. This specialization and trade was sponsored through elites who forged ties to the wider region with Ancestral Puebloans, perhaps to mitigate food shortages and reduce conflict and warfare. During the thirteenth century, the Ancestral Puebloans faced a severe "demographic upheaval" (read: social crisis that forced depopulation and migration) that brought migrants to the Mogollon and Hohokam regions.

Bayman (2001) writes that the Post-Classic period 1450–1540 CE, "witnessed an enigmatic termination of Hohokam society shortly before or immediately after European contact in North America." Unfortunately, there is a gap of coverage and understanding about the timing and cause of collapse, and Bayman recites all of the following variables that have been argued to cause the collapse:

1 Floods
2 Droughts
3 Internal/local disease
4 Disease from European contact
5 Internal warfare
6 External warfare/invasion
7 Soil degradation
8 Canal erosion
9 Earthquakes.

While there is evidence and strong support particularly for the role of floods during the time of collapse, Bayman writes that a "growing number of archeologists are dissatisfied with predominately ecological explanations of precontact change in the North American Southwest."

Indeed, anthropologist Michael Wilcox (2010) believes that this crisis was initiated through European contact, and that there has been a deep misunderstanding about both the Hohokam and others regional groups, such as the Ancestral Puebloans (formally referred to as Anasazi) in Chaco Canyon. This interpretation of European cause of collapse is consistent with what other scholars on the Hohokam have found as well, though there is still debate about the timing and cause of settlement abandonments in the Sonora.

The Hohokam are probably the ancestors of the Yuman and the O'odham (formally referred to as Pima), but there are strong linguistic and cultural ties to contemporary Hopi and Zuni tribes.

Wilcox warns that we have fundamentally misunderstood radical changes in both the Sonoran Desert and in Chaco Canyon, and that the rise and fall of these cultures have wrongly been written as a myth of disappearance and collapse due to ecological malfeasance on the part of the tribes—to the exclusion of the impact of empire. Wilcox argues that the abandonment of sites of the Hohokam were purposeful and that the Hohokam moved to different sites to avoid the violent coercion of the Spanish.

Is it possible that the Hohokam had established one of the most sustainable societies (some *dwellings* lasted far longer than the age of the US republic) but faced radical social change and collapse because of European imperial ambitions? In this chapter we explore the complex causes of social and civilization collapse, which again, appears to be consistent with phases of the Adaptive Cycle. However, there is a debate currently in anthropology about the nature of these collapses.

CHECKPOINT: DO CIVILIZATIONS *REALLY* COLLAPSE?

While it is clear that the Hohokam did face a collapse in that the society fragmented and lost it social organization as well as suffering a depopulation event, the O'odham and other likely descendants of the Hohokam continue to survive in the Sonoran Desert to this day. If the ancestors of the Hohokam still live in the region, but at the same time the Hohokam is a storied collapse, do collapses really happen? Perhaps, a better question is—"do collapses occur the way they have been imagined in popular culture?" Perhaps because of religious traditions that foresee total apocalypse, there is an ongoing notion that collapse means that everyone in the population dies. This is not the kind of collapse evident in the historical record. Indeed, there is a danger in oversimplifying how collapse has occurred and thus it is important to read the relevant literature in anthropology, archeology, and history carefully because these are the major disciplines that rigorously study these events and processes.

Some commentators on collapse, such as famed biologist Jared Diamond (2005), have argued that environmental problems in conjunction with political choices have led to the collapse of many civilizations. However, in a review of the topic, Lawler (2010) quotes important subtleties that appeal to what Trinity College Dublin historian Poul Holm calls an "industry of apocalypse that pervades religion, academia, and even Hollywood, with its blockbusters like *2012* [a film about global apocalypse that popular culture believed was predicted by the Mayan calendar]." Since Diamond's work on collapses, anthropologists have warned against **environmental determinism**, where environmental conditions determine human conditions. Environmental determinism is opposite end to **human exemptionalism**, where environmental conditions are completely separate and do not affect human conditions at all (Chapter 3), as a spectrum of beliefs.

Part of the problem comes from the difficulty of speaking across different scholarly traditions, where physical scientists locate a physical change, such as drought, but may gloss over what we know about social systems and social resilience. Under this condition, environmental change becomes a simple, "snappy explanation" but one without a good sense of how people adapt (Holm in Lawler, 2010).

A few lessons that scholarship in this area reveal:

1 Collapses are not caused by simple or singular reasons.
2 Populations under stress move, and so even if one part of a society collapses and loses significant population, other parts of that civilization may adapt or migrate out of the situation, where they then recreate their culture and ways of life. For example, while the Lowland Maya certainly collapsed and were almost totally depopulated, other areas of Mayan culture survived and continues to do so even today.
3 Civilization stresses are geographically uneven, and what happens in one region does not necessarily occur in others. Thus, collapse scholarship must be as specific as possible about the regional variations.
4 There are few, perhaps no, cases of collapse where everyone dies. As in #2, migration is a logical response to severe pressures, and parts of former civilizations live on.
5 Collapse is a systems concept, and one kind of collapse, such as collapse of a political regime, may not initiate whole civilization collapse; therefore it is important to be clear about what systems we are referring to when discussing collapse—i.e., collapse of what?

Also, the term "collapse" carries some ambiguity. Cambridge scholar Colin Renfrew describes collapse as "the loss of central administration, disappearance of an elite, decline in settlements, and a loss of social and political complexity" but the concept itself is difficult to strictly define (paraphrased by Lawler, 2010). Joseph Tainter, one of the foremost authorities in civilization collapse, defines **collapse** similarly:

> Collapse, as viewed in the present work, is a political process. It may, and often does, have consequences in such areas as economics, art, and literature, but it is fundamentally a matter of the sociopolitical sphere. A society has collapsed when it displays a rapid, significant loss of an established level of sociopolitical complexity. The term "established level" is important. To qualify as an instance of collapse a society must have been at, or developing toward, a level of complexity for more than one or two generations.
>
> (Tainter, 1988)

Thus, collapse implies an abrupt change, but how abrupt? If we are discussing a civilization that had existed for several thousand years, if its population declines over a hundred years that is "abrupt" relative to the life of the society, so some conditions of collapse are relative to other points of reference.

If we mean that a system has lost its complexity and is simplified abruptly compared to how long it took to build that complexity, then systems collapse all the time, and Tainter makes it clear that complex civilizations have collapsed *as a regularity across human history.* We can also refer back to systems thinking, and speak of collapse as a regime shift in social–ecological systems (Beddoe et al., 2009). If, however, we take collapse to mean, "the complete end of those political systems and their accompanying civilizational framework" then it rarely, if ever, occurs (Eisenstadt quoted in McAnany and Yoffee, 2010).

This chapter will therefore focus on the notion of systemic loss of complexity across different systems in order to think about collapse.

HOW DO COLLAPSES OCCUR?

Bert M. de Vries (2007) writes that the academic literature on collapse shows three primary types of collapse. These types are problems that come from the society or that come from outside the society and are not able to be handled.

1 Resource and environment related changes, fully external or partly internally induced changes
2 Interaction-related changes in the form of conquest or other, less dramatic forms of invasion
3 Internal changes in sociopolitical, cultural, and religious organization and world-view that diminish the adequacy of response to external events.

External environmental changes have typically been climatic in nature, seen in climatic variations of warmer or colder periods that affect food harvest and water availability. For example, Weiss and Bradley (2001) note that "The Akkadian empire of Mesopotamia, the pyramid-constructing Old Kingdom civilization of Egypt, the Harappan 3B civilization of the Indus valley, and the Early Bronze III civilizations of Palestine, Greece, and Crete all reached their economic peak at about 2300 B.C." And then they were hit with a "catastrophic drought and cooling that generated regional abandonment, collapse, and habitat-tracking [moving with habitat changes]" (Weiss and Bradley, 2001). Thus, they find that, "Many lines of evidence now point to climate forcing as the primary agent in repeated social collapse" (Ibid).

This history raises important considerations for sustaining 9–11 billion people on agricultural yields that are dependent on stable climatic conditions and has Weiss and Bradley concerned for future global stability. However, while climate does affect civilization collapses in the most sophisticated models, the intersection of climatic and social change is more complicated because social strategies can build resilience against these problems, more noted on this below, p. 183 (Butzer, 2012).

In addition to climatic conditions, internal changes also affect societies. Internal environmental changes can build suitable conditions for the transmission of diseases and exact deforestation, erosion of soil, and loss of biodiversity through hunting, etc. Changes in demographics, usually a rise in population, are also critical when population overshoots the ability of the land to produce enough food and confounds the society's ability to take care of itself in other ways.

Naturally, all the reasons we have for civilization collapses fall into these major types but these causes usually intersect and overlap in specific cases, as in the fall of Rome. Rome suffered internal sociopolitical changes that made it susceptible to the Visigoths that invaded. Usually we hear that Rome fell because the invading Germanic tribes overran Rome. This can be thought of as the proximate cause, or the straw that broke the camel's back, so to speak. However, the Romans had been vanquishing the Goths for perhaps 1000 years. Consequently, that one simple cause does not adequately explain the collapse of Rome, just around 500 CE. Instead, it appears that Rome first weakened itself, and this is the ultimate cause, or root cause of the collapse of Rome.

In fact, Rome had been weakening at the turn of the century, and nearly collapsed 250 years earlier. Rome had solved some of its problems earlier by feeding a rising population through conquest and expansion, essentially creating granaries out of other regions like northern Africa. At first, this strategy worked marvelously, for Rome anyway; it worked less well for northern Africa. In addition to a rising population, free grain was offered to peasants and the Roman nobility as political bribery to satisfy political problems. To help pay for this, the empire devalued its currency, and this caused rampant inflation. Over time, the Roman dinar was essentially worthless, and on the backside of these solutions was a larger problem that ultimately spelled the death of the empire.

COLLAPSE OF COMPLEX CIVILIZATIONS

Tainter defines complex societies as those that have social hierarchical inequality and functional differences in a society (e.g., specialized divisions of labor). He notes that complex societies tend to "expand and dominate," and, "today they control most of the earth's lands and people." Tainter plumbs the archeological record to understand the collapse of complex civilizations, such the Mayan and Roman empires, the Harappan Civilization, Mesopotamia, the Hittite Empire, and Minoan Civilization among others. Consistent with all the writers highlighted in this section, Tainter sees collapse as political—that means it is about choice and is manifest in social change.

Tainter (1988) outlines several principal theories of civilization collapse, and first among them is "depletion or cessation of a vital resource or resources on which the society depends." This theory posits that a civilization is subject to

> gradual deterioration or depletion of a resource base (usually agriculture), often due to human mismanagement, and the more rapid loss of resources due to an environmental fluctuation or climatic shift. Both are thought to cause collapse through depletion of the resources on which a complex society depends.

Tainter rejects this this theory because it assumes that civilizations sit idly by as their future slips away, but complex civilizations are designed to solve exactly these problems. Thus, he rejects simple resource exhaustion theory of civilization collapse because complex human organization is designed to avoid this basic civilization problem. Instead, Tainter believes that complex civilizations collapse ultimately because of their own complexity.

He argues that when a civilization encounters a problem, like a food shortage, it tends to solve it with more complexity (hierarchy, bureaucracy, spending, and specialization). But, this complexity comes with a cost—say the cost of new trades/skills to solve new problems, or expanding and conquering new lands. At some point in this process, the cost of complex solutions becomes too marginal to solve the problem. This cost is paid through an energy subsidy, which is where a society feeds more energy it obtains from somewhere or someone else into the system to cover these increasing costs. Naturally, this energy comes from resources and ecological systems as well as people. So, Tainter rejects the idea that societies collapse from simple resource scarcity, but rather as a result of resource scarcity that comes from complex problem solving. We might then summerize Tainter's ideas of where collapse comes from as:

- a failure of social problem solving;
- resulting in a shortage of energy to maintain the complex society.

Unfortunately, also, once complexity is used to solve a problem, it is not a simple matter to scale it back. Think of how modern countries solve problems sometimes by adding bureaucracy. Each addition in bureaucratic problem solving costs energy and resources, but it is no small matter to eliminate the same layers. For example, the simple human terms of hiring someone and then having to lay them off creates a cost that is either born at the government level, or in aggregate at the social level as people suffer from unemployment and its attendant problems. Also, bureaucratic organizations take funding and governmental commitments, and administrators defend these lines of funding and authority with vigor—for example, the US Department of Defense is extremely effective at defending against cuts to its budget and authority. In his farewell address, the US President Eisenhower warned about this military-industrial complex becoming so powerful that it threatened the ability of leaders like Presidents to make decisions without the military-industrial complex's consent, similar to Tainter's warning:

> A complex society pursuing the expansion option, if it is successful, ultimately reaches a point where further expansion requires too high a marginal cost. Linear miles of border to be defended, size of area to be administered, size of the required administration, internal pacification costs, travel distance between the capital and the frontier, and the presence of competitors combine to exert a depressing effect on further growth. ... Once conquered, subject lands and their populations must be controlled, administered and defended ... . Ultimately the marginal returns for the conquest start to fall, whereupon the society is back to its previous predicament.
>
> (Tainter, 1988)

This is actually what happened to Rome, according to Tainter. The Roman imperial expansion fed Roman citizens initially but, at some point, the energy needed to maintain these imperial projects became too much, and conquests eventually only paid enough to feed the invading soldiers. Thus, the ultimate cause of the collapse of Rome, Tainter argues, was complexity itself because there is a limited ability for this approach to keep solving Rome's problems, which ultimately came down to feeding and pacifying its own citizens while defending against outside threats. Consequently, even as a society pours more energy into solving its problems, that energy becomes less and less capable of making a difference.

In reference to our contemporary globalized world, Tainter is unsure if industrial society writ large has reached the point where the marginal return of energy investment is in decline, but that it *"inevitably will"*—"In ancient societies the solution to declining marginal returns was to capture a new energy subsidy. In economic systems activated largely by agriculture, livestock and human labor (and ultimately by solar energy), this was accomplished by territorial expansion" as in Rome and the Ch'in of Warring States China, as well as "countless empire builders" (Tainter, 1988). But, "In an economy that today is activated by stored energy reserves, and especially in a world that is full, this course is not feasible (nor was it ever permanently successful) … . A new energy subsidy will at some point be essential."

Further, as a connected world industrial society, "Collapse, if and when it comes again, will this time be global. No longer can any individual nation collapse. World civilization will disintegrate as a whole. Competitors who evolve as peers collapse in like manner."

ADAPTIVE CYCLE, PANARCHY, AND COLLAPSE

Through Tainter's work, we know that collapse involves complex systems, and the most advanced theories of collapse work from that starting point. As a set of complex systems we know that collapse does not occur from one single cause, but from a train of events that cascade out from triggers of decline in complexity that are consistent and follow the **Adaptive Cycle** within a **Panarchy**.

These triggers are conditioned by environmental degradation and climatic changes that can spur food shortages that can lead to sub-sistence crises, rural flight and urban decline, and epidemics that can lead a civilization open to demolition of the social order, civil and foreign war, depopulation, political fragmentation, and foreign domination. In this train of events, there are pathways for more resilient responses and responses that foster further breakdown.

"The process of breakdown typically begins with economic or fiscal decline caused by external and internal inputs" serving as preconditions for collapse (Butzer, 2012). The economic decline can lead to adaptive or maladaptive actions that stabilize or accel-erate collapse. The threshold, or tipping point toward collapse, is lowered by environmental degradation—e.g., groundwater deple-tion, salinization of soil, soil erosion, deforestation, and soil degra-dation reduce productivity and can lead to food shortages. Climatic changes, then, accelerate environmental degradation and can "unleash more catastrophic forms" of extreme events and natural disasters like drought or flooding.

When lower economic productivity interrupts trade food commodities while **ecosystem services** are damaged and dampen local food production or exposing people to infectious diseases, the stage is set for a subsistence crisis where the basic needs of people are not being met, and depopulation worms its way across the civilization.

THE FALL OF THE LOWLAND MAYA

The Maya lowlands stretched from the current Mexican Yucatan Peninsula to Belize, Guatemala, and Honduras (see Figure 7.1).

The civilization was complex, building remarkable ceremonial centers in Tikal and Copan, establishing powerful city-states, developing nuanced understanding of their environment while

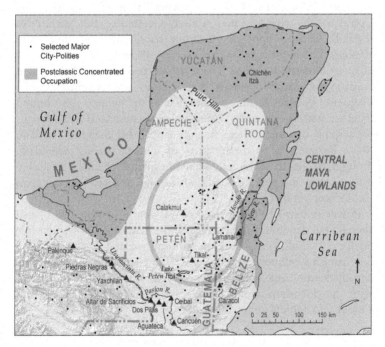

Figure 7.1 Classic Lowland Maya realm

creating intensive economies with long-distance trade of prestige goods (e.g., jade ornaments) and subsistence goods (foods). They occupied the Lowlands for an astonishing 2000+ years prior to their collapse. The civilization peaked between 700 and 800 CE, though areas like Chitzan Itza flourished afterward, perhaps as a sign that their political and military ambitions contributed to the decline of other areas. One of the better studied areas is the urban center of Copan, which had pyramids, tombs, temples, and complex artistic features like sculptures. During its peak, this city probably had 10–12,000 people per square kilometer. However, by 1200 CE, the population was only around 1000 in total.

In the field of anthropology, the fall of the classic Lowland Maya is a continuing mystery that has been examined through decades of vast research, and knowledge about the Lowland Maya continues to expand. For example, while Caracol, Belize has been studied for decades by anthropologists, new advanced mapping techniques expanded the area of study by 800 percent just in 2011 (Chase et al.). Consequently, what we know is surely incomplete. But, within this research, there do appear to be some consistent conclusions even amongst the vigorous debate: the Maya collapsed because of a series of inter-connected stresses. These stresses occurred in the coupled human and natural systems that follows the Adaptive Cycle. Ecological changes on the ground and in the climate were bound together with social, cultural, and political conditions that brought on a spiral of crises. There is wide agreement that the Lowland Maya decline was a combination of climatic stress, particularly drought, interacted with other environmental changes, including deforestation, and political conditions that led to social fragmentation and violent conflict.

Power centers of the Lowland Maya were abandoned and faced almost total depopulation in areas like the central Petén District (see Figure 7.1). Still, since some areas flourished and it was not until the Spanish Conquest that the pre-Columbian Maya met their systematic demise, some scholars have been cautious for reasons consistent with the discussion above about using the word "collapse"—but Turner and Sabloff write:

> The Central Maya Lowlands (CMLs) ... and its large infrastructure of cities, water systems, and managed landscapes, were essentially

abandoned, however, with population declines approaching 90%, and it remained so for well over a millennium. In this sense, the term collapse is appropriate.

(Turner and Sabloff, 2012)

Studies of Mayan skeletons and bone show that over time, nutrition and health declined. Bones become shorter over time; enamel of children's teeth degraded. Rebecca Storey (1992) found that the teeth in particular indicate that children experienced substantial nutritional stress around age three, which is the typical age of weaning, in Copan at the time of the collapse. She discovered that even children buried in privileged groups were exposed to the stress of nutritional and infectious disease. However, the archeological and anthropological research is clear: the Maya collapse cannot be viewed from simple environmental problems, but rather as coupled social-ecological causes of decline.

Mayan specialists Nicholas Dunning and colleagues (2012) have shown that the Maya collapse probably experienced a fatal "risk spiral" that was set in place starting with deforestation, that correlated with two different collapses, the Terminal Pre-Classic of the second century CE and the Terminal Classic of the ninth and tenth centuries. These risk spirals were not spread uniformly around the entire Lowland region, because abandoned settlements and depopulation occurred through a patchwork that followed trends of exploitation and growth, collapse, and reorganization.

The vulnerabilities of different urban centers depended on their geographic location and their sociopolitical adaptations. For example, within sections of the interior Lowland on the Yucatan is a set of raised plateaus and basins of the Elevated Interior Region (EIR) with rainfall that varies based on latitude, and areas in the EIR have less access to freshwater sources. This makes these areas more vulnerable to drought. Drought is not just categorized by lack of meteorological precipitation, but also has three other manifestations relevant to the Maya: agricultural drought, which reduces soil moisture through farming and land clearing; hydrological drought that reduces surface water levels in streams and lakes; and socio-economic drought, which disrupts water sources by producing water-dependent goods and services. These forms of drought link together where, for example, meteorological drought can lead to agricultural drought.

Deforestation reduces transpiration from forest vegetation reducing precipitation that exacerbates drought conditions and reduces soil moisture. Soil is regenerated through ash and phosphorous (lacking in Lowland soils but necessary for agriculture), but soil regeneration is slowed when the tree canopy that catches the ash and phosphorus is absent. As the needs of the Maya increase with higher population, the fallow (rest and recovery of the forest) decreases, and this reduces nitrogen in the soil, decreasing soil fertility. This change in soil opens up the area to weed invasions, and cropped plants like maize become more vulnerable to disease. Here, the good original conditions for agriculture led to "the incremental reduction of forest for construction material, fuel, and farm land associated with population growth and urbanization would have created a risk spiral within the region, especially when coupled with other environmental and cultural risk factors" (Dunning, Beach, and Luzzadder-Beach, 2012). These increasing vulnerabilities *also* linked with sociopolitical conditions where "Maya polities thrived on growth and expansion that funneled wealth to a small ruling elite," that eventually weakened the social structure that would become destabilized by warfare (Ibid). "In short, system rigidity and poor options for change created conditions ripe for collapse" (Ibid). Other scholars agree: "a very small percent of the population … maintained significant authority and power, with major disparities in wealth and standard of living over the vast majority of the population," which was needed for fabled scales of labor, "required to managed forests and opened lands, capture and retain water, reclaim wetlands, sustain monumental building projects, and fill the ranks of the military to combat and raid other city-states, all during a time of increasing aridity" (Turner and Sabloff, 2012)

However, amongst the disparity of experiences within the region, at times there can be sociopolitical collapse that does not reduce population on the same level. Indeed, one aspect of society can rapidly lose complexity while other elements remain intact—a political system can collapse without depopulation (US invasion of Iraq and the subsequent collapse of the Hussein regime did not lead to the collapse of the Iraqi ancient culture, nor systematic depopulation). Likewise, economic systems can collapse to be replaced without a collapse in the cultural or political systems. These systems are separated out in Table 7.1, which provides a rough approximation of initial variables associated with the collapse of the respective system indicated. A sideways figure-eight is used in Table 7.1 to indicate that these variables are interrelated and themselves have their own systemic conditions, and sometimes the variable may even be a cause or an effect of collapse in the respective system (Butzer, 2012). Also, it should be said that because a variable is related to the collapse of a system type, it does not ensure that such collapse occurs; however, crisis in one system affects the other systems—the mechanisms through which crises spread or do not

Table 7.1 Variables for collapse across key social systems

Variables of collapse	Socio-cultural system	Political system	Economic system	Population collapse
Technological and capital change	∞		∞	
Religious and symbolic changes	∞			
Linguistic change	∞			
Cultural memory or knowledge	∞		∞	
Inflation			∞	
Entitlements (commitments to pay a particular group)		∞	∞	
Corruption		∞		
Loss of political unification		∞		
Civil war	∞	∞		∞
Foreign war	∞	∞		∞
Famine	∞	∞		∞
Food shortage			∞	
Disease				∞
Drought			∞	
Flooding			∞	
Water chemistry changes			∞	∞
Soil fertility			∞	∞
Deforestation			∞	∞
Climatic temperature changes			∞	

spread from one system to another is determined by the severity of the crisis and how tightly the systems are connected.

One determining factor for collapse across systems is the way in which variables trigger or feed into a train of other conditions—a resilient society will be able to reduce or reverse negative changes. Other societies will be more vulnerable to crossing a threshold or breaking point that sets in motion a downward spiral of inter-connected crises, seen in the Back Loop of the Adaptive Cycle. Collapse of one part of a social system may not mean ruin for the civilization, but if the civilization is caught up in a risk spiral, often with increasing population, changes to climate or soil fertility can bring that society closer to a threshold of rapid and unwanted change. Also, if we look at the category with the most variables that are related to collapse, collapse of the economic system has the most. The economic system in place for any society is part of how subsistence occurs, which can lead to political and population col-lapse as well. Indeed, Butzer finds that decline usually begins in the economic system, but that social and political systems have been the most fragile component, at least for ancient states.

DARK AGES

Like the collapse of civilizations, **Dark Ages** are crises of a system—but, they are a decline of whole **world-systems**. World-systems, a concept first discussed by Wallerstein (1974–80), are single multi-state economies with a division of labor, and they are referred to as "world" systems because the system covers an area larger than the largest political territory of that time. They are divided by the labor of core (central) powers, periphery (marginalized powers), and semi-periphery states which share qualities of the latter and the former. Since that time, research has dramatically expanded world-systems theory to also include the single systems during world history characterized by:

1 Bulk commodities and primary goods (Wallerstein's model)
2 Prestige goods networks
3 Military networks
4 Information-based networks.

The central idea, however, since Wallerstein's main contribution, is that there are now and have been organized single social networks through space and time that cross and connect societies; and, that these systems will be affected by each other as energy, material, and information are exchanged. The world-system we currently live in is connected globally with different intensities and timing of 1–4. Crisis in one society in the system affects others, but with a lag time and different effects in different areas of the system. Also, as noted at the beginning of this chapter, there are disparate privileges in world-systems, and crisis will be seen by elites and others as harmful, but in the periphery it may be seen as an opportunity to re-organize.

Dark Ages, according to scholar Sing Chew, "are periods, in some cases, of contraction and/or collapses of human communities and civilisations," which "exhibit conditions of acute social, economic, and political disruptions exhibited by trends such as economic slowdowns, structural social and political breakdowns, de-urbanisation, increased/reduced migration, and population losses" (Chew, 2002).

Dark Ages include collapse of civilizations but are treated differently here because:

1 The scope of Dark Ages are at the world-systems level (both ecological and social), whereas civilization collapses refer to singular societies.
2 Dark Ages include the aftermath of civilization collapse, whereas theories of collapse tend to focus most on the causes and moment of collapse.

As world-systems contain feedbacks, many have internal relationships and can self-organize, they are complex adaptive systems that are characterized by non-linear cycles. All world-systems pulsate, or expand and decline, at various speeds.

Chew's analysis of Dark Ages and ecology come from an empirical examination of ecological degradation within the various civilizations he studied spanning 5000 years and comparing the onset of ecological degradation and the Dark Age that these societies experienced. His first insight is that:

Over world history, the relationship between culture and nature has been punctuated with periods of ecological degradation and crisis. Given these outcomes, the history of human civilisations can therefore also be described as the "history of ecological degradation and crisis." ... In this regard, Dark Ages are at times exhibiting ecological degradation, climatic changes, reorganisation of socio-economic and political structures, and hegemonic challenges.

(Chew, 2002)

Part of this systemic crisis is a vast loss of cultural memory, including basic skills of pottery and writing. We can see that one Dark Age comes on the heels of the collapse of Greek society. V. R. Desborough in *The Greek Dark Ages* describes this period as such:

during these generations the changes that came about are little short of fantastic. The craftsmen and artists seem to vanish almost without a trace: there is very little new stone construction of any sort, far less any massive edifices; the metal-worker's technique reverts to primitive, and the potter, except in early stages, loses his purpose and inspiration; and the art of writing is forgotten. But the outstanding feature is that by the end of the twelfth century the population appears to have dwindled to about one-tenth of what it had been a little over a century before.

(Desborough, 1972)

Whereas periods of social expansion tend to favor more economic intensity and exploitive technologies, during Dark Ages many of the techniques and technologies that supported expanding social and economic growth are lost. Since expert divisions of labor control the production of specific things, like making barrels or wheels, when the social system that supports these divisions falls apart, few people are left with the skills to make wheels for themselves because they did not have to do it before. If we were in the midst of a Dark Age now, for example, few people would know how to build a mobile phone. Thus, after the loss of expertise, the larger population is unable to complete basic functions that were commonplace prior to the crisis.

Historical Dark Ages typically take at least 500 years to end and allow social reconstruction. Dark Ages are long phases of economic regress, and potential phases of ecological regeneration. There is historic evidence that the health of economic growth and environmental quality grow counter-posed to each other that are difficult to refute, and economic intensity appears (over the long duration, and, perhaps even shorter ones) to be inversely proportional to ecological health.

That being said, *because the economic intensity triggers some of the ecological crisis*, Dark Ages are initially characterized by ecological crisis at the beginning and middle phases, and even sometimes the entire duration of the Dark Age. If ecological crisis continues through the entire duration, then the Dark Age affords less regenerative opportunities for ecological systems.

Included in the ecological crisis, common problems are "deforestation levels, devastated landscapes, soil erosion, and endangered species underlining these periods" (Chew, 2007). It also is the case the climatic variations often overlap with these other problems, exaggerating the crisis. Note that until the Industrial Revolution when social action began to move coal and other hydrocarbons out of the ground, climate change was an outside factor not caused by people but climatic change certainly affected people's prospects.

Chew explains,

> Climatological changes are also associated with Dark Ages. Climatological changes and natural calamities, when they occur during Dark Ages generate further challenges to social system reproduction ... . Higher-than-normal temperatures can generate salinization problems for agricultural cultivation, especially in areas where irrigation is extensively used. It could also lower harvest yields. The aridity that commonly occurs with high temperatures has often generated severe problems for pastoral herds because of the loss of foliage and grasses that have led to nomadic migrations, thus causing further social pressures on core centers [the powerful urban centers of the region].
>
> (Chew, 2007)

DARK AGES IN SUM

Dark Ages are triggered and experienced during and after civilization collapses. They are systemic and involve serious losses of technology, knowledge, skills, and population. Dark Ages, similar to civilization collapses, are connected and caused by economic activity that pushes the ecological system into crisis, throwing the social system into crisis. They are often punctuated by climate variations. Sometimes Dark Ages allow for regeneration of ecological systems, and Chew warns that this then obscures the connections of ecology and cultural pressures for the next social system, who then may commit the same Normative Failure of failing to limit ecological degradation. These periods, it should be obvious, also allow for a re-arrangement of power. If these are system-wide crises, and they eviscerate the prior social system, after at least 500 years, there is an opening for a new arrangement of power.

Hegemonic cycles/sequence are cycles in the world-system where core hegemons (most powerful actors in the system) dominate the system and its organization. Secular cycles/sequence are cycles where the world-system is breaking down, and the hegemony of the core powers are in decline, usually along with population numbers. Dark Ages are an example of a secular cycle.

How and why do world-systems cycle? Hall and Turchin believe that these oscillations come down to the political reactions to loss of agricultural surplus and population. One series of connections they propose is that:

> population growth in excess of the productivity gains of the land leads to state fiscal crisis because of persistent price inflation. This in turn leads to expansion of armies and rising real costs [of living]. Population growth also leads to an increased number of aspirants for elite positions, which puts further fiscal strains on the state, leading to increased intra-elite competition, rivalry, and factionalism. These strains on the state lead to popular discontent because of falling real wages, rural misery, and urban migration. Intensification of these trends eventually causes state bankruptcy and consequent loss of military control, giving way to elite movements of regional and national rebellion.
>
> (Hall and Turchin, 2007)

The study of these oscillations are very complex, but "Historically, the cycles of expansion lead to long-term degradation of larger areas and then in turn to the transformation of the social order" (Friedman, 2007). Given the reality that world-systems are linked, these cycles themselves spread over the connected geographies.

Thus, the first world-system of the Bronze Age, inclusive of Harrappa, Mesopotamia, and Ancient Egypt, experienced a Dark Age following the collapse of the world-system at the time and the geography of this crisis covered the area below, which lasted from 2200–1700 BC This Dark Age comes on the heels of the first stage in the Agricultural Revolution and closes out the Bronze Age between the three major civilizations, which together made up a world-system that experienced a synchronized secular sequence. Remember that pre-historic civilization contains the Neolithic (Stone Age), then the Bronze Age, and then the Iron Age—though sometimes the Copper Age is part of this, and the sequence is not always the same in each civilization.

After the Bronze Age suffered its Dark Age and that world was able to recover, economic connections were more intensive, and the world-system continued to grow. In the following world-system, the Greeks were in control of the system.

Of course, since world-systems always "pulse," for the Dark Age following this second Dark Age, the geographic scale is noticeably larger. In the third episode of a Dark Age, the pattern of expansion widens even further. It is because of this trend that Chew (2007, 2002), Ponting (2007), and others agree to some extent that the *next* Dark Age will also follow the geographic pattern of the world-system, but today that world-system is, unlike any prior time, virtually the entire planet, and thus a Dark Age in a fully globalized world-system would probably encompass the globe.

WHAT DO WE KNOW?

First, all of these theories for collapse have different nuances to them, but they all include environmental change in some way and maladaptive political causes in some way. Thus, simple resource scarcity appears to be just as unlikely to cause collapse as simple invasion, or one bad leader. We can conclude:

1 Civilizations do not collapse for simple (single) causal reasons. Problems overlap and interact.
2 Ecological conditions underlie the actual pressures that make people move (emigrate) or perish. Often this has to do with agricultural surplus and the growth of populations beyond the region's ability to feed a society.

3 Political regimes are pivotal in all cases because civilizations have choices about causing and solving problems. Politics is a key to survival, and a reason for demise. Note, however, that because of these political factors, Malthus was not entirely correct—he thought (at least in his early writing) that it was inevitable that people outgrow their food supply, but there are choices all along the way to such misery.

From the histories of collapse we continue to see the importance of the problem structure of sustainability, First Principles, and the nature of Normative Failure. The fact that ecological conditions underlie the very basic requirements of keeping a society going means that theories of collapse verify First Principles. Indeed, the maintenance of ecological systems and their ability to provide basic life supports (food, water, absorption of poisons, etc.) are necessary but insufficient conditions for the continuity of a civilization. Normative—political—structures are the more contested social issues of ethics, governments, and accountability that make up the harder facts of sustainability and its opposite, collapse. *Collapse, therefore, is the result of Normative Failure.*

Further, we know that when societies intensify and expand their economic activity, it eventually leads to a systemic crisis, or Dark Age, that takes at least 500 years to recover from. This crisis includes a failure of memory, and the origin and causes of the Normative Failure are often forgotten by the next generation who commit the same errors again. In addition, we know that Dark Ages occur in linked economic systems, and over time, economic world-systems have progressively grown in scope to the current planetary scale. Dark Age crises occur over the whole scale of the respective world-system. In today's world, few would be exempt from such a crisis. Finally, all world-systems pulsate—expansion and degradation. These cycles recur throughout history, and everyone knows it is foolish to annoy history by ignoring it.

CRITICAL CONSIDERATIONS

How do you think the lessons of past collapse apply to the modern era? What do you think the role of hierarchy, ethics, and justice are in the process of civilization collapse?

Some scholars warn that the world is about to enter into a Dark Age. What do you think the criteria for measuring a global Dark Age would be, and what do you think the consequences would be? What do you think are the most important Normative Failures that lead to collapse from the ones listed in this chapter?

WHAT DO YOU THINK OF THESE SUSTAINABILITY SOLUTIONS?

Would any of these help avert a modern collapse?

1   Closing energy loops in urban areas where waste is captured as a resource.
2   Agriculture shifts to more agro-ecological models that mimic natural cycles and systems so that erosion, biocides, and pollution are minimized.
3   Elimination of non-essential "consumptive" (where the water does not go back into the natural water cycle appropriately). water use.
4   Any major project must listen and defer to those most affected, and for those who cannot be at any discussion, like future generations or non-humans, the decision is made "as if" they were there and their interests were important considerations.
5   Certain things are deemed illegal to privatize, and declared the common heritage of humankind for all: water, the gene pool, seeds, the atmosphere.

WHAT DO YOU THINK OF THE FOLLOWING SYLLOGISM?

Premise A: Dark Ages across time have covered the world-system (the interconnected market at the time).

Premise B: The world-system covers the entire planet at this point in history.

Conclusion: The next Dark Age, should it occur, would geographically include the entire planet and all its inhabitants.

FURTHER READING

Hornberg, Alf and Carole Crumley. (2007). *The World System and the Earth System: Global Socioenvironmental Change and Sustainability Since the Neolithic.*

Walnut Creek, CA: Left Coast Press. This book is a rigorous review of Dark Ages and major documented cycles in world-systems across time.

Wallerstein, Immanuel. (1989). *The Modern World System*. New York: Academic Press. Wallerstein is the original scholar of world-systems theory, and this is an easy to read synopsis.

McAnany, Patricia. A. and Norman Yoffee. (2010). *Questioning Collapse: Human Resilience, Ecological Vulnerability, and the Aftermath of Empire*. Cambridge, UK: Cambridge University Press. This book takes issue with the concept of overly simplistic or dramatic descriptions of collapse and provides a counter to environmental determinism.

Tainter, Joseph A. (1988). *The Collapse of Complex Societies*. Cambridge, UK: Cambridge University Press. While collapse theories are controversial, Tainter's book and proposed theory, that complexity starts as a solution but cannot be maintained over time driving collapse, remain a classic and hold up well to the latest thinking.

Ponting, Clive. (2007). *A New Green History of the World: The Environment and the Collapse of Great Civilizations*. London: Penguin. A very accessible but critical account of the big movements across human history that explain vast inequalities and threats to sustainability at large scales.

Jacobs, Jane. (2010). *Dark Age Ahead*. New York: Random House. One of the great urban thinkers, Jacobs explains her fear that important social boundaries like accountability are being lost and that we will face a Dark Age without correcting these issues.

CONCLUSION

The First Principles of Sustainability indicate it is the process of building and maintaining global social systems of capable, accountable, adaptive, just, and free people who can make important decisions and trade-offs with foresight and prudence who foster the robust, self-organizing, dynamic, and complex ecosystems around the world for now and future generations.

Thought of this way, it is possible that, while there are many stories of failure, the ongoing struggles to make this world better and to keep the imperfect human prospect alive is the larger story-line of sustainability written on the tablets of human history. We can see some of the subsidiary storylines play out across this text.

There are severe contradictions in pursuing one kind of welfare that can come at the expense of other sources of welfare. Our economic welfare, or more to the point, the economic welfare of some, may come at the expense of social equity or ecological health; and, while highlighting social equity or ecological health may compromise economic growth, all of these conditions depend on complex interdependence, where "we can never do just one thing" because whatever actions we take in the world we affect a million other things across time (Thiele, 2011).

However, we are in fact forced to act in one way or another, and our assumptions of scarcity and abundance, optimism or pessimism about the future affect what we actually choose to do. If we assume, as the British did, that famine of the poor is inevitable, we might choose to do nothing in the face of severe death and misery. If we assume, on the other hand, that there are no true problems of sustainability, such as limits to critical Adaptive Cycles, again we may choose to do nothing. Instead, we may be best off to envision our ability to solve any problem, but to take our problems as real and important and worthy of solving. However, the discussion about measuring sustainability tells us that the nature of these challenges will require both objective and subjective measures, and this problem makes the essentially contested nature of sustainability an even larger challenge to govern.

Another storyline is that the decisions we make in this complex and interdependent world will have distributive effects on the life chances of people around the world. This means that the way we negotiate the broader international moral order relates to life and death, survival and continuity, and sustainability and oblivion. However, we should expect that these moral struggles for survival among other problems will continue to host power struggles through the classic arenas of domination that will affect current and future generations. Unfortunately, as societies establish themselves, a hubris tends to blind leaders and their people to the history of collapse and Dark Ages that have worked their away into history as a regularity.

While it is very difficult to positively identity sustainability in complex social-ecological systems, it is easier to identify when societies are not sustainable because they reduce economic welfare, increase inequalities, and degrade ecological systems. Some have argued that the story of the human species will come to a conclusion if we do not keep a "reasonable" balance:

> It is not so much that humanity is trying to sustain the natural world, but rather that humanity is trying to sustain itself. It is us that will have to "go" unless we can put the world around us in reasonable order. The precariousness of nature is *our* peril, *our* fragility.
>
> (Sen, 2013, emphasis in original)

Empires, war, globalization, and state-led economic growth all have spread the economic benefits and ecological harms unevenly, while some 60 percent of the world's ecosystem services have been degraded (Millennium Ecosystem Assessment, 2005b). Imagine how long it took to build and establish these systems.

We have seen that there are important reasons for hope, when we see something like Engineers Without Borders (EWB) use simple, inexpensive technologies to open opportunities for self-determination and sustainable infrastructures that save lives. For example, when the US EWB founder in the USA, Bernard Amadei, addressed engineering students at the University of Central Florida in October 2011, he noted the following example of what we can do. He described how he was visiting Kabul, Afghanistan. In Afghanistan it is cold and there is a shortage of fuel (remember our ethics discussion about energy in Chapter 3). There are also a lot of international agencies working in Kabul that have a lot of shredded paper from their bureaucratic functions which they dispose of without a thought. Amadei and his colleagues were able to take $20 worth of iron and make a simple crank press that when fed paper, made fuel disks. These disks could be sold for a price everyday Afghans could afford. This crank was given to a legless man who had been conscribed to begging for food, who then had meaningful employment. They then found four boys to sell the disks on the street during half of the day. These boys were child prostitutes beforehand, but now spent half the day making an income, and the other half they could attend school. The emphasis of design and engineering, when driven by a responsible ethic, can provide sustainable solutions that save lives and improve the welfare of the worst off. Inside the city, these people were living lives that were not empowered, but the opportunities for honest work through inexpensive but durable design made a difference. I suspect that the metaphor most relevant to this situation was one of Spaceship Earth first described by Kenneth Boulding, where the materials we have and the systems on Earth are fragile and we must think about them deliberately and with justice to live sustainably.

Opposite Spaceship Earth is what Boulding called "Cowboy Economics" consistent with a metaphor of infinite and timeless frontiers with infinitely stable ecological systems that can never be exhausted or disturbed. This kind of metaphor leads to design, engineering, and political projects that often look like the Belo

Monte Dam or the Alberta Tar Sands that serve a minority of people while undermining social and ecological integrity.

Each of these storylines make up a larger discussion about how people will live on this planet, what future generations' prospects are, and the nature even of the rest of life on Earth in the Anthropocene.

It is my personal hope that we begin to change the basic operating principles of world civilization to be more consistent with Bernard Amadei and Engineers without Borders, Satprem Maïni and the Auroville Institute, and the late Wangari Maathai and the Green Belt Movement than those of the tar sands, Belo Monte Dam or the Environmental Skepticism/Climate Denial projects. I sincerely believe that our children's children will share this preference.

FURTHER READING

Litfin, Karen. (2014) *Eco-Villages: Lessons for Sustainable Community*. Cambridge, UK: Polity Press. Litfin beautifully weaves together stories of people and communities, both in affluent and poor countries, who have constructed sustainable and good lives together.

Maniates, Michael and John M. Meyer. (2010). *The Environmental Politics of Sacrifice*. Cambridge, MA: MIT Press. So much about sustainability sounds like sacrifice; these authors reshape the way we think about sacrifice, in part by recognizing what some people already do sacrifice for a good life.

Davis, Wade. (2009). *The Wayfinders: Why Ancient Wisdom Matters in the Modern World*. Toronto, Canada: House of Anansi. Diverse culture and memory are a key to our survival, Davis elegantly chronicles this drama.

Ridgeway, Sharon and Peter Jacques. (2013). *The Power of the Talking Stick: Indigenous Politics and the World Ecological Crisis*. Boulder, CO: Paradigm Publishers. In this book, co-author Sharon Ridgeway and I lay a critique of the current world system and ask that the world indigenous movement be heard. Leaders in this movement argue unapologetically that we must defend Mother Earth if humanity is to have a future.

GLOSSARY

Adaptive Cycle: A model of how complex systems change over time, and it links ecosystems and social systems in "never-ending adaptive cycles of growth, accumulation, restructuring, and renewal" (Holling, 2001).

Adaptive governance: Making decisions to building resilient social-ecological systems to make societies less vulnerable and more resilient to the dangerous prospects of rapid changes at the local level that are expected when slow changes are made at the global level.

Agency: Having the ability to make choices.

Agricultural Revolution: The move to sedentary agriculture from hunter-gatherer subsistence that began with emmer wheat planted in the Fertile Crescent some 10,000 years ago.

Anthropocene: The period in geological history where human activity dominates every ecosystem on Earth.

Anthropocentrism: A value set that places humans at the peak of global importance. **Humanistic anthropocentrism** believes protecting nature is important because it will help other people; **deep anthropocentrism** holds no or very little value for nature at all.

Arenas of domination: That political space where state, corporate, and civil society projects vie for influence not just over the nuts-and-bolts of policy, but over the broad moral landscape in which policy is made.

Autonomy (civil society): When civil society can express authentic interests of its own, and these interests are not dictated from the state of corporations.

Biocentrism: Values all life.

Biodiversity: "the sum total of all of the plants, animals, fungi, and microorganisms on Earth; their genetic and phenotypic variation; and the communities and ecosystems of which they are a part" (Dirzo and Raven, 2003).

Capital: An asset that produces income or surplus; can be human, financial, physical, social, or ecological/natural capital. These assets produce the basis of all consumption.

Carbon cycle: The vast cycle of mobilization and uptake of carbon through the Earth's terrestrial, marine, and atmospheric systems.

Carrying Capacity: The maximum number of individuals that can be sustained in a population without harming the ecosystems they depend upon.

Civil society: That part of society that is neither in charge of state governments nor large corporations and economic production; the association of people who make up non-governmental organizations, social movements, and other non-state and non-corporate organizations.

Collapse: The sudden and dramatic loss of complexity.

Collective action problems: Problems that arise when individuals resist cooperating with each other in order to provide a common good for everyone.

Corporate Social Responsibility (CSR): When businesses authentically take responsibility for the impacts of business.

Countermovement: A conscious, organized attempt to oppose the claims and efforts of a social movement, typically to reduce or remove changes demanded by the social movement.

Dark Ages: Periods of broad social crisis across world-systems that follow the collapse of single civilizations.

Demographic transition: The theory that pre-industrial groups have high fertility and mortality rates and slow population growth. During industrialization, health and nutrition improve, increasing lifespans of the population, spurring a fast population growth rate. As populations become more secure they have fewer children and population growth rates stabilize or decline.

The Difference Principle: A proposal from John Rawls that if a rational person did not know what kinds of privileges or obstacles s/he had in life and therefore s/he did not know what kinds of resources—rights, wealth, opportunities, etc.—they possessed either, that person would agree to rules that only allowed for inequality when it benefited the least well off.

Division of Labor: when different people hold different, specialized jobs to fulfill social needs.

Ecocentrism: Values everything within ecological systems including landscapes.

Ecological Footprint: A measure of consumption translated into the amount of land or sea needed for that consumption.

Economism: When economic conditions are valued by people more than other values, such as social or ecological values.

Ecosystem services: Natural capital; ecological goods and services that are critical for human welfare. Ecosystem services come in four types: provisioning, cultural, regulating, and supporting.

Environmental determinism: The mistaken assumption that environmental conditions totally determine what actions people take.

Environmental Skepticism/Climate Denial Countermovement: The proposition that (mostly) global environmental problems, such as climate change, are inauthentic—they are not real and/or important, so there are no problems with the long term-continuity of the human species. Environmental and climate denial is deployed by a conservative countermovement in an attempt to protect modern Western progress and Northern consumptive habits and power.

Epistemology: The study, process, or framework of how we know what we think we might know. Often discussed as "knowledge systems."

Essentially contested terms: An essentially contested term is one that has several legitimate meanings that cannot be resolved through argumentation.

Exponential growth: Growth by a fraction of the stock over time at a constant rate.

First Principles of Sustainability: Core requirements for the long-term continuity of any society:

P1: Without ecological life supports, there is no society. This relationship is immutable. A sustainable society must maintain the integrity of Earth systems and cycles that provide critical life supports.

P2: What kind of society that grows in an ecological space is a value-based question, but sustainable societies must observe normative constraints:

1. The social system will not be sustainable if it undermines ecological life supports (principle of accountability and restraint)
2. The social system will not be sustainable if it sufficiently militates against itself or is annihilated by others (principle of justice).
3. The social system must be adaptive to challenges and changes to avoid evolving vulnerabilities (principle of foresight).

Free rider problem: A **collective action problem** where an individual benefits from a publicly provided resource without paying.

Green Revolution: The project started in the 1940s to industrialize agriculture with chemical inputs, high-yield varieties of crops, and eventually genetic modification for higher yield per acre agriculture.

Hegemony: Power that the people do not generally question, but that embodies rules of society that are normalized to actually reinforce the interests of the ruling class through the state.

Heuristics: Rules of thumb that come from trial and error. Heuristics can be most helpful when precise measures are unavailable or too expensive to gather, though heuristics can also lead to incorrect generalizations.

Human exemptionalism: The set of beliefs that humanity is so distinct and special that it is exempt from the rest of the laws of nature, ecological limits, or even evolutionary pressures, including extinction.

Industrial Revolution: Started in Great Britain around 250 years ago, industrialization involves switching from solar energy held in plants to solar energy held in coal and oil (and at first wood) to power machinery such as the steam engine. With the harnessing of more power and energy, and hence more work, such as larger cropped fields, the Industrial Revolution facilitated the growth of large economies and major social transformations from changes in warfare to changes in household dynamics and work.

Institutional fitness: When institutional rules fit the environmental problem and scale.

Institutional interplay: When institutions affect each other.

Institutions: Systems of rules, norms, strategies, roles, and decision-making procedures; institutions can be formal laws, or informal expectations, or anywhere in-between. Institutions are not organizations, even though organizations may have and enforce rules.

Life chances: The opportunities an individual has to improve the quality and duration of her/his life.

Limits to Growth (LTG): In addition to being a series of books led by Donella Meadows, the limits to growth is a concept regarding the constraints of any system.

Maximum Sustainable Yield (MSY): A theoretical amount of harvest that can be removed from a resource, like the amount of fish or timber, without damaging the population.

Million species–years: A way to calculate extinction rates, species-years refers to every year every species is expected to exist, where if there are 15 million species, every year is 15 million species-years and each species has an average range of 1 to 10 million years of existence. The normal or background extinction rate is 1–0.1 species per million species–years or 0.0001 percent, understood by measuring extinctions through the fossil record; the current extinction rate is estimated to be closer to 0.1 percent of estimated species per year, 1000 times greater than the rate without human disturbance to the web of life.

Moral Standing: Deserving recognition for inherent value.

Neoliberalism: A political economic system that attempts to shift power from state and social arenas to economic arenas; it favors deregulation and reduced state spending on social safety nets and environmental protections.

Net primary production (NPP): Primary production is the rate of biomass growth that comes from plant respiration and photo-synthesis, where gross primary production is the sum total of all the energy produced by plants but some of this energy is used for plant growth. The remaining energy produced but not used by the plant is net primary production.

Non–human person: An entity endowed with the value and recognition of having its own interests as a person, but that is not human.

Non–renewable resource: Resources, such as oil, that do not reproduce or renew over time. **Renewable resources**, such as fisheries, forests, and most aquifers, reproduce or renew themselves over time.

Normative Failure: Failure of a society or group of societies to institute the normative injunctions against P1 requirements, and therefore they fail to provide the normative constraints to ensure basic needs.

Ontology: "Onto" means being, and an ontology is way of being, life way, and sense of human purpose in the world.

Overshoot: The overuse of available resources, but still provides a time for correction before a collapse.

Panarchy: Nested, hierarchical adaptive cycles.

Planetary Boundaries: The estimated preconditions for human welfare.

Popular sovereignty: The power of civil society to convert its majority demands into policy.

Precautionary Principle (PP): The PP indicates that protective action should not wait for scientific certainty, and that when making an environmentally risky decision we should err on the side of caution and restraint to protect human health, future ecosystem services like medicines, research, the life of non-humans, and other non-economic values.

Problem structure of sustainability: The problem that all organisms have a metabolism that requires consumption and disposal of energy and matter provided by ecological systems; but, consumption disturbs the very ecosystems necessary for a healthy metabolism to begin with and therefore the life of the organisms over time.

Regime shifts—aka catastrophic shifts or state changes: Rapid changes of the system from one stable state to a different state.

Resilience: The capacity of a system to experience a disturbance, and then return to its original state, avoiding a regime shift.

Scales: Linked systems across dimensions of time and space.

Sixth Great Extinction: The current pace of biodiversity loss is 100–1000 times normal extinction rates, and because there have been five other punctuated periods of extinction in geologic time, this is the Sixth Great Extinction.

Social capital: Social networks of knowledge, trust, institutions, and systems of reciprocity.

Social-ecological system: The coupled or integrated social and ecological systems that co-evolve.

Social Movement: A conscious and organized attempt to make demands on society, typically from grassroots efforts. See also: **countermovement**.

Sovereignty: The power to control the interests of a political group, usually a country, with rights of non-interference from others.

Structural ecological changes: Changes to ecosystems that alter the larger systems and not just parts of these systems, such as to biological communities, nutrient and chemical cycles, the climate system, hydrologic cycle, and other ecological system-wide changes.

The Structure of Political Opportunity: Contextual conditions regarding the capacity and organization of nation-states, critical events that focus public attention and concern, and the arrangement of interests of different groups, including the availability of alliances that allow or restrict opportunities for civil society influence.

Sustainability (see debate on defining sustainability in Chapter 2): Literal definition: to hold up; Brundtland definition of "sustainable development" is development that "meets the needs of the present without compromising the ability of future generations to meet their own needs"; the author's definition: the imperfect process of building and maintaining global social systems of capable, accountable, adaptive, just, and free people who can make important decisions and trade-offs with foresight and prudence who foster the robust, self-organizing, dynamic, and complex ecosystems around the world for now and future generations.

Synergistic: Dynamic interaction of effects.

System: An organized set of parts that create a larger unified whole that neither one of the parts could have produced alone. A **complex system** is one that has many internal parts and many relationships between these parts, so that changing one part produces mostly unpredictable results.

Three Es: The so-called three pillars of sustainability are social equity, economic health, and ecological integrity.

Threshold: A place where slow mounting changes build up and "deep uncertainty explodes," creating massive change (Holling, 2003).

Throughput: The economic activity of "take, make, and waste" from and into natural systems, consuming ecosystem capacities.

Tragedy of the commons: A kind of **collective action problem** where individuals have an incentive to undermine the common good for their own interests and there are few rules to prevent them from doing so.

Triple Bottom Line (TBL): A qualitative accounting framework for measuring sustainable progress through people, profits, and planetary improvement.

Vulnerability: The opposite of resilience, vulnerability is the susceptibility to disturbances. Vulnerability varies based on the nature of the system and the exposure and sensitivity to a particular threat.

Weak versus strong sustainability: Weak sustainability argues that we can continue to consume ecological goods and services at a growing rate, whereas strong sustainability insists on strict limits to consuming ecosystems that means more radical social changes are required for sustainability under the strong version.

World Ocean: The ensemble of connected basins of the Atlantic, Pacific, Indian, Southern, and Arctic Oceans.

World-systems: Single multi-state economies with a division of labor that cover an area larger than the largest political territory of that time. They are divided by the labor of core (central) powers, periphery (marginalized powers) and semi-periphery states that share qualities of the latter and the former.

REFERENCES

Adger, W. N., Brown, K. and Tompkins, E. L. 2005. The political economy of cross-scale networks in resource co-management. *Ecology & Society*, 10(2). Available: www.ecologyandsociety.org/vol10/iss2/art9. Accessed 5/28/2014.

Alam, D. S., Chowdhury, M. A. H., Siddiquee, A. T., Ahmed, S., Hossain, M. D., Pervin, S., Streatfield, K., Cravioto, A. and Niessen, L. W. 2012. Adult cardiopulmonary mortality and indoor air pollution: A 10-year retrospective cohort study in a low-income rural setting. *Global Heart*, 7(3), 215–21.

Anand, S. and Sen, A. 2000. Human development and economic sustainability. *World Development*, 28, 2029–49.

Anderegg, W. R. L., Berry, J. A., Smith, D. D., Sperry, J. S., Anderegg, L. D. L. and Field, C. B. 2012. The roles of hydraulic and carbon stress in a widespread climate-induced forest die-off. *Proceedings of the National Academy of Sciences*, 109, 233–37.

Andersson, K. and Agrawal, A. 2011. Inequalities, institutions, and forest commons. *Global Environmental Change*, 21(3), 866–75.

Angel, J. L. 1984. Health as a crucial factor in the changes from hunting to developed farming in the eastern Mediterranean. In: Cohen, M. N. and Armelagos, G. J. (eds.) *Paleopathology at the Origins of Agriculture*. (pp. 51–74). London: Academic Press.

Archer, D., Eby, M., Brovkin, V., Ridgwell, A., Cao, L., Mikolajewicz, U., Caldeira, K., Matsumoto, K., Munhoven, G., Montenegro, A. and Tokos, K. 2009. Atmospheric lifetime of fossil fuel carbon dioxide. *Annual Review of Earth and Planetary Sciences*, 37, 117–34.

Arrow, K., Dasgupta, P., Goulder, L., Daily, G., Ehrlich, P., Heal, G., Levin, S., Mäler, K.-G. R., Stephen Schneider, Starrett, D. and Walker, B. 2004. Are we consuming too much? *Journal of Economic Perspectives*, 18, 147–72.

AtKisson, A. 2012. *Believing Cassandra: How to be an Optimist in a Pessimist's World*. 2nd ed. Hoboken, NJ: Earthscan.

Auroville Earth Institute. 2013. *Welcome to Earth Architecture!* [Online]. Auroville, Tami Nadu, India. Available: www.earth-auroville.com/index. php. Accessed 11/24/2013.

Bagchi, A. K. 2005. *Perilous Passage: Mankind and the Global Ascendancy of Capital*. Lanham, MD: Rowman & Littlefield.

Baker, N. D. 1933. The "new spirit" and its critics. *Foreign Affairs*, 12, 1–19.

Barrett, C. B., Brandon, K., Gibson, C. and Gjertsen, H. 2001. Conserving tropical biodiversity amid weak institutions. *Bioscience*, 51, 497–502.

Bayman, J. M. 2001. The Hohokam of southwest North America. *Journal of World Prehistory*, 15, 257–311.

Beck, U. 1999. *World Risk Society*. Cambridge, UK: Polity Press.

Beckerman, W. 2002. *A Poverty of Reason: Sustainable Development and Economic Growth*. Oakland, CA: Independent Institute.

Beddoe, R., Costanza, R., Farley, J., Garza, E., Kent, J., Kubiszewski, I., Martinez, L., McCowen, T., Murphy, K., Myers, N., Ogden, Z., Stapleton, K. and Woodward, J. 2009. Overcoming systemic roadblocks to sustainability: The evolutionary redesign of worldviews, institutions, and technologies. *Proceedings of the National Academy of Sciences*, 106, 2483–89.

Bettencourt, L. M. A. and Kaur, J. 2011. Evolution and structure of sustainability science. *Proceedings of the National Academy of Sciences*, 108(49), 19540–45.

Biermann, F., Abbott, K., Andresen, S., Bäckstrand, K., Bernstein, S., Betsill, M. M., Bulkeley, H., Cashore, B., Clapp, J. and Folke, C. 2012a. Navigating the Anthropocene: Improving earth system governance. *Science*, 335, 1306–7.

Biermann, F., Abbott, K., Andresen, S., Bäckstrand, K., Bernstein, S., Betsill, M. M., Bulkeley, H., Cashore, B., Clapp, J. and Folke, C. 2012b. Transforming governance and institutions for global sustainability: Key insights from the Earth System Governance Project. *Current Opinion in Environmental Sustainability*, 4, 51–60.

Blomqvist, L., Brook, B. W., Ellis, E. C., Kareiva, P. M., Nordhaus, T. and Shellenberger, M. 2013. The Ecological Footprint remains a misleading metric of global sustainability. *PLoS Biology*, 11, e1001702.

Borucke, M., Galli, A., Iha, K., Lazarus, E., Mattoon, S., Morales, J. C., Poblete, P. and Wackernagel, M. 2013. *The National Footprints Account 2012 Edition* [Online]. Oakland, CA: Global Footprint Network. Available: www. footprintnetwork.org/images/article_uploads/National_Footprint_Accounts_ 2012_Edition_Report.pdf. Accessed 11/26/2013.

Boserup, E. 2005. *The Conditions of Agricultural Growth: The Economics of Agrarian Change under Population Pressure*. New Brunswick; London: Aldine Transaction.

Boulanger, P.-M. 2011. The life-chances concept: A sociological perspective in equity and sustainable development. In: Rauschmayer, F., Omann, I. and Fruhmann, J. (eds.) *Sustainable Development: Capabilities, Needs, and Well-being*. Abingdon, Oxon, UK: Routledge.

Brondizio, E. S., Ostrom, E. and Young, O. R. 2009. Connectivity and the governance of multilevel social-ecological systems: The role of social capital. *Annual Review of Environment and Resources*, 34, 253–78.

Buck, S. 1985. No tragedy of the commons. *Environmental Ethics*, 7, 48–54.

Bulletin of the Atomic Scientists. 2007a. Doomsday Clock Overview. http://thebulletin.org. Accessed 5/28/2014.

Bulletin of the Atomic Scientists. 2007b. It is 5 Minutes to Midnight. http://thebulletin.org. Accessed 5/28/2014.

Butzer, K. W. 2012. Collapse, environment, and society. *Proceedings of the National Academy of Sciences*, 109, 3632–39.

Cardinale, B. J., Duffy, J. E., Gonzalez, A., Hooper, D. U., Perrings, C., Venail, P., Narwani, A., Mace, G. M., Tilman, D. and Wardle, D. A. 2012. Biodiversity loss and its impact on humanity. *Nature*, 486, 59–67.

Carpenter, P. A. and Bishop, P. C. 2009. The seventh mass extinction: Human-caused events contribute to a fatal consequence. *Futures*, 41, 715–22.

Casebeer, W. D. 2003. Moral cognition and its neural constituents. *Nature Reviews Neuroscience*, 4, 840–45.

Cassman, K. G., Matson, P. A., Naylor, R. and Polasky, S. 2002. Agricultural sustainability and intensive production practices. *Nature*, 418, 671–77.

Centeno, M. A. and Cohen, J. N. 2012. The arc of neoliberalism. *Annual Review of Sociology*, 38, 317–40.

Chapin III, F. S., Zavaleta, E. S., Eviner, V. T., Naylor, R. L., Vitousek, P. M., Reynolds, H. L., Hooper, D. U., Lavorel, S., Sala, O. E., Hobbie, S. E., Mack, M. C. and Diaz, S. 2000. Consequences of changing biodiversity. *Nature*, 405, 234–42.

Chase, A. F., Chase, D. Z., Weishampel, J. F., Drake, J. B., Shrestha, R. L., Slatton, K. C., Awe, J. J. and Carter, W. E. 2011. Airborne LiDAR, archaeology, and the ancient Maya landscape at Caracol, Belize. *Journal of Archaeological Science*, 38, 387–98.

Chew, S. 2002. Globalisation, ecological crisis, and Dark Ages. *Global Society*, 16, 333–56.

Chew, S. 2007. *The Recurring Dark Ages: Ecological Stress, Climate Changes, and System Transformation*. Lanham, MD: AltaMira Press/Rowman & Littlefield.

Clark, M. 1999. Fisheries for orange roughy on seamounts in New Zealand. *Oceanologica Acta*, 22(6), 593–602.

Communications Directorate Fisheries and Oceans. 1993. *Charting a New Course: Towards the Fishery of the Future: Report of the Task Force on Incomes and Adjustment in the Atlantic Fishery*. Ottowa, Ontario.

Corson, C. and MacDonald, K. I. 2012. Enclosing the global commons: The convention on biological diversity and green grabbing. *Journal of Peasant Studies*, 39, 263–83.

Daly, H. E. 2011. *Ecuadorian Court Recognizes Constitutional Right to Nature* [Online]. Chester, PA: Widener Environmental Law Center Blog. Available: http://blogs.law.widener.edu/envirolawblog/2011/07/12/ecuadorian-court-recognizes-constitutional-right-to-nature/. Accessed 4/20/2013.

Daly, H., Czech, B., Trauger, D. L., Rees, W. E., Grover, M., Dobson, T. and Trombulak, S. C. 2006. Are we consuming too much—for what? *Conservation Biology*, 21, 1359–62.

Davidson, D. J. and Gismondi, M. 2011. *Challenging Legitimacy at the Precipice of Energy Calamity*. New York: Springer.

Davison, A. 2008. Contesting sustainability in theory-practice: In praise of ambivalence. *Continuum: Journal of Media & Cultural Studies*, 22, 191–99.

De Tocqueville, A. 2003. *Democracy in America and Two Essays on America*. New York: Penguin Classic.

De Vries, B. 2007. In search of sustainability: What can we learn from the past? In: Hornborg, A. and Crumely, C. (eds.) *The World System and the Earth System: Global Socioenvironmental Change and Sustainability since the Neolithic*. Walnut Creek, CA: Left Coast Press.

De Vries, B. and Goudsblom, J. (eds.). 2003. *Mappae Mundi: Humans and Their Habitats in a Long-Term Socio-Ecological Perspective: Myths, Maps and Models*. Amsterdam: Amsterdam University Press.

De Waal, F. 2005. *Our Inner Ape: A Leading Primatologist Explains Why We Are Who We Are*. New York: Riverhead Books.

De Waal, F. B. M. 2008. Putting the altruism back into altruism: The evolution of empathy. *Annual Review of Psychology*, 59, 279–300.

Desborough, V. R. D. A. 1972. *The Greek Dark Ages*, London: Benn.

DeSombre, E. R. 2006. Global environmental institutions. In: Weiss T. and Wilkinson, R. (eds.) *Global Institutions*. Abingdon, Oxon, UK: Routledge.

Desormeaux, D., Jenson, D. and Enz, M. K. 2005. The first of the (Black) memorialists: Toussaint Louverture. *Yale French Studies*, 107, 131–45.

Diamond, J. 2005. *Collapse: How Societies Choose to Fail or Succeed*. New York, NY: Viking/Allen Lane.

Diaz, R. J. and Rosenberg, R. 2008. Spreading dead zones and consequences for marine ecosystems. *Science*, 321, 926–29.

Dietz, T., Fitzgerald, A. and Shwom, R. 2005. Environmental values. *Annual Review of Environment and Resources*, 30, 335–72.

Dirzo, R. and Raven, P. H. 2003. Global state of biodiversity and loss. *Annual Review of Environment and Resources*, 28, 137–67.

Discovery Institute. 2012. *About Discovery* [Online]. Seattle, WA. Available: www.discovery.org/about.php. Accessed 12/4/2012.

Dobson, A. 2000. *Green Political Thought*. New York: Routledge.

Dobson, A. 2003. *Citizenship and the Environment*. Oxford: Oxford University Press.

Dower, N. 2004. Global economy, justice and sustainability. *Ethical Theory and Moral Practice*, 7, 399–415.

Doyle, T. 1998. Sustainable development and Agenda 21: The secular bible of global free markets and pluralist democracy. *Third World Quarterly*, 19, 771–86.

Dresner, S. 2008. *The Principles of Sustainability* (second edn.). London: Earthscan.

Dunlap, R. E. 2002. Paradigms, theories, and environmental sociology. In: Dunlap, R. E., Buttel, F. H., Dickens, P. and Gijswijt, A. (eds.) *Sociological Theory and the Environment: Classical Foundations, Contemporary Insights*. (pp. 329–50). Lanham, MD: Rowman & Littlefield.

Dunlap, R. E. and Van Liere, K. D. 1984. Commitment to the dominant social paradigm and concern for environmental quality. *Social Science Quarterly*, 65, 1013–28.

Dunning, N. P., Beach, T. P. and Luzzadder-Beach, S. 2012. Kax and kol: Collapse and resilience in lowland Maya civilization. *Proceedings of the National Academy of Sciences*, 109, 3652–57.

Emerson, J. W., Hsu, A., Levy, M. A., Sherbinin, A. D., Mara, V., Esty, D. C. and Jaiteh, M. 2012. *2012 Environmental Performance Index and Pilot Trend Environmental Performance Index*. New Haven, CT: Yale Center for Environmental Law and Policy.

Ezzati, M. and Kammen, D. M. 2002. Household energy, indoor air pollution, and health in developing countries: Knowledge base for effective interventions. *Annual Review of Energy and the Environment*, 27(1), 233–70.

Ferguson, N. 2010. Complexity and collapse. *Foreign Affairs*, 89(2).

Ferguson, N. 2011. *Civilization: The Six Ways the West Beat the Rest*. New York: Allen Lane.

Fiala, N. 2008. Measuring sustainability: Why the ecological footprint is bad economics and bad environmental science. *Ecological Economics*, 67, 519–25.

Fischer, J., Manning, A. D., Steffen, W., Rose, D. B., Daniell, K., Felton, A., Garnett, S., Gilna, B., Heinsohn, R., Lindenmayer, D. B., Macdonald, B., Mills, F., Newell, B., Reid, J., Robin, L., Sherren, K. and Wade, A. 2007. Mind the sustainability gap. *Trends in Ecology & Evolution*, 22(12), 621–24.

Fish, S. K. and Fish, P. R. 2008. *The Hohokam Millennium*. Santa Fe, NM: School for Advanced Research Press.

Folke, C. 2006. Resilience: The emergence of a perspective for social–ecological systems analyses. *Global Environmental Change*, 16, 253–67.

Folke, C. and Kåberger, T. 1991. Recent trends in linking the natural Environment and the economy. In: Folke, C. and Kåberger, T. (eds.) *Linking the Natural Environment and the Economy: Essays from the Eco-Eco Group.* Dordrecht: Kluwer.

Folke, C., Hahn, T., Olsson, P. and Norberg, J. 2005. Adaptive governance of social-ecological systems. *Annual Review of Environment and Resources*, 30, 441–73.

Friedman, J. 2007. Sustainable unsustainability: Toward a comparative study of hegemonic decline in global systems. In: Hornborg, A. and Crumley, C. (eds.) *The World System and the Earth System: Global Socio-environmental Change and Sustainability Since the Neolithic.* Walnut Creek, CA: Left Coast Press.

Friedman, H. and McMichael, P. 1989. Agriculture and the state system: The rise and decline of national agricultures, 1870 to the present. *Sociologia Ruralis*, 29(2), 93–117. doi:10.1111/j.1467-9523.1989.tb00360.x

Fuller, R. B. 2008. *Operating Manual for Spaceship Earth.* Zürich: Lars Müller Publishers.

Gaodi, X., Shuyan, C., Qisen, Y., Xia Lin, F., Zhiyong, Boping, C., Shuang, Z., Youde, C., Liqiang, G., Cook, S. and Humphrey, S. 2012. *China Ecological Footprint 2012: Consumption, Production, and Sustainable Development.* Beijing: World Wildlife Fund Beijing Office, Global Footprint Network, Institute of Zoology, Zoological Society of London, China Council for International Cooperation on Environment and Development.

Garcia, S. M. and Rosenberg, A. A. 2010. Food security and marine capture fisheries: Characteristics, trends, drivers and future perspectives. *Philosophical Transactions of the Royal Society B: Biological Sciences*, 365, 2869–80.

Gareau, B. 2013. *From Precaution to Profit: Contemporary Challenges to Environmental Protection in the Montreal Protocol.* New Haven, CT: Yale University Press.

Gibbon, E. 1994. *The History of the Decline and Fall of the Roman Empire.* London/New York: Penguin.

Gilligan, C. 1995. Hearing the difference: Theorizing connection. *Hypatia*, 10, 120–27.

Gillis, J. 2011. With deaths of forests, a loss of key climate protectors. *The New York Times* 1 October, p. A1.

Godwin, W. 1798. *Enquiry Concerning Political Justice, and Its Influence on Morals and Happiness: By William Godwin.* London: G. G. and J. Robinson.

Goodall, J. 1998. Essays on science and society: Learning from the chimpanzees: A message humans can understand. *Science*, 282(5397), 2184–85.

Goodall, J. 2003. Bridging the chasm: Helping people and the environment across Africa. *Environmental Change & Security Project Report*, 9, 1–5.

Goodland, R. 1995. The concept of sustainability. *Annual Review of Ecology and Systematics*, 26, 1–24.

Gramsci, A. 1996. *Prison Notebooks*. Edited by J. A. Buttigieg. Vol. 1–3. New York: Columbia University Press.

Haldane, A. 2012. The dog and the frisbee. *Federal Reserve Bank of Kansas City's 36th economic policy symposium, "The Changing Policy Landscape"*, Jackson Hole, Wyoming.

Hale, S. 2010. The new politics of climate change: Why we are failing and how we will succeed. *Environmental Politics*, 19, 255–75.

Hall, A. and Branford, S. 2012. Development, dams and Dilma: The saga of Belo Monte. *Critical Sociology*, 38, 851–62.

Hall, T. and Turchin, P. 2007. Lessons from population ecology for world-systems analyses of long-distance synchrony. In: Hornberg, A. and Crumely, C. (eds.) *The World System and the Earth System*. Walnut Creek, CA: Left Coast Press.

Hande, H. H. 2007. Reliable, renewable rural energy. In: C. E. Smith *Design for the Other 90%*. C. (pp. 47–49). New York: Cooper-Hewitt, National Design Museum/Smithsonian Institute.

Hansen, J., Sato, M., Kharecha, P., Beerling, D., Berner, R., Masson-Delmotte, V., Pagani, M., Raymo, M., Royer, D. L. and Zachos, J. C. 2008. Target atmospheric CO_2: Where should humanity aim? *The Open Atmospheric Science Journal*, 2.

Hardin, G. 1968. The tragedy of the commons. *Science*, 162, 43–48.

Hardin, G. 1974. Commentary: Living on a lifeboat. *Bioscience*, 24(10), 561–68.

Hardin, G. 1995. *Living within Limits: Ecology, Economics, and Population Taboos*: New York: Oxford University Press.

Hawken, P., Lovins, A. and Lovins, L. H. 1999. *Natural Capitalism: Creating the Next Industrial Revolution*. New York, Little, Brown and Company.

Hawken, P., Lovins, A. B. and Lovins, L. H. 2010. *Natural Capitalism: The Next Industrial Revolution*. London: Earthscan Publications.

Hay, P. R. 2002. *Main Currents in Western Environmental Thought*. Bloomington, IN: Indiana University Press.

Hayanga, A. J. 2006. Wangari Mathai: An African woman's environmental and geopolitical landscape. *International Journal of Environmental Studies*, 63, 551–55.

Heather, P. 2006. *The Fall of the Roman Empire: A New History of Rome and the Barbarians*. Oxford: Oxford University Press.

Hempel, L. C. 2009. Conceptual and analytical challenges in building sustainable communities. In: Mazmanian, D. A. and Kraft, M. E. (eds.) *Toward Sustainable Communities: Transition and Transformations in Environmental Policy*. Cambridge, MA: MIT Press.

Hertel, T. W. 2011. The global supply and demand for agricultural land in 2050: A perfect storm in the making? *American Journal of Agricultural Economics*, 93, 259–75.

Hoekstra, A. Y. and Mekonnen, M. M. 2012. The water footprint of humanity. *Proceedings of the National Academy of Sciences*, 109, 3232–37.

Höhler, S. 2010. The environment as a life support system: The case of Biosphere 2. *History and Technology*, 26, 39–58.

Holling, C. S. 2001. Understanding the complexity of economic, ecological, and social systems. *Ecosystems*, 4, 390–405.

Holling, C. S. 2003. *Foreword: The Backloop to Sustainability.* Cambridge: Cambridge University Press.

Holloway, G. and Sou, T. 2002. Has Arctic sea ice rapidly thinned? *Journal of Climate*, 15, 1691–1701.

Huber, P. 1999. *Hard Green: Saving the Environment from the Environmentalists, A Conservative Manifesto.* New York, NY: Basic Books.

Hull, R. 2005. All about Eve: A report on environmental virtue ethics today. *Ethics & the Environment*, 10, 89–110.

Humphreys, D. 2006. *Logjam: Deforestation and the Crisis of Global Governance.* London: Earthscan.

Huseman, J. and Short, D. 2012. "A slow industrial genocide": Tar sands and the indigenous peoples of northern Alberta. *The International Journal of Human Rights*, 16, 216–37.

Indigenous Peoples Global Conference on Rio+20 and Mother Earth. 2012. *Kari-Oca 2 Declaration.* Rio de Janeiro, Brazil.

International Indigenous Peoples Summit on Sustainable Development. 2002. *The Kimberaly Declaration of the International Indigenous Peoples Summit on Sustainable Development* [Online]. Available: www.iwgia.org/sw217.asp.

IPCC. 2013. Climate change 2013: The physical science basis. In: Stocker, T. F., Qin, D., Plattner, G.-K., Tignor, M., Allen, S. K., Boschung, J., Nauels, A., Xia, Y., Bex, V. and Midgley, P. M. (eds.). Cambridge, UK: Working Group I to the Fifth Assessment Report of the Intergovernmental Panel on Climate Change.

Jackson, J. B. C. 2008. Ecological extinction and evolution in the brave new ocean. *Proceedings of the National Academy of Sciences*, 105, 11458–65.

Jacobs, M. 1999. Sustainable development as a contested concept. In: Dobson, A. (ed.) *Fairness and Futurity.* Oxford: Oxford University Press.

Jacques, P. J. 2009. *Environmental Skepticism: Ecology, Power, and Public Life.* Burlington, VT; Surrey, UK: Ashgate Publishing Ltd.

Jacques, P. J. 2014. Emerging issues: Civil society in an environmental context. In: Fairfax, S. and Russell, E. (eds.) *The Guide to US Environmental Policy.* Washington, DC: CQ Press.

Jacques, P. J., Dunlap, R. E. and Freeman, M. 2008. The organization of denial: Conservative think tanks and environmental scepticism. *Environmental Politics*, 17, 349–85.

Jaenicke-Despres, V., Buckler, E. S., Smith, B. D., Gilbert, M. T. P., Cooper, A., Doebley, J. and Pääbo, S. 2003. Early allelic selection in maize as revealed by ancient DNA. *Science*, 302, 1206–8.

Jenni, K. 2005. Western environmental ethics: An overview. *Journal of Chinese Philosophy*, 32, 1–17.

Jones, A. H. M. 1964. *The Later Roman Empire, 284–602: A Social, Economic, and Administrative Survey*. Baltimore, MD: Johns Hopkins University Press.

Jones, R. C. 2012. Science, sentience, and animal welfare. *Biology & Philosophy*, 28(1), 1–30.

Kanninen, T. 2013. *Crisis of Global Sustainability*. London: Routledge.

Kari-Oca Declaration. 1992. Kari-Oca Declaration.

Kates, R. W., Parris, T. M. and Leiserowitz, A. A. 2005. What is sustainable development? Goals, indicators, values, and practice. *Environment (Washington DC)*, 47, 8–21.

Kertcher, C. 2012. From Cold War to a system of peacekeeping operations: The discussions on peacekeeping operations in the UN during the 1980s up to 1992. *Journal of Contemporary History*, 47, 611–37.

Kirch, P. V. 2005. Archaeology and global change: The holocene record. *Annual Review of Environment and Resources*, 30, 409–40.

Korsgaard, C. M. 2004. Fellow creatures: Kantian ethics and our duties to animals. *Tanner Lecture on Human Values* University of Michigan.

Kummu, M., Ward, P. J., de Moel, H. and Varis, O. 2010. Is physical water scarcity a new phenomenon? Global assessment of water shortage over the last two millennia. *Environmental Research Letters*, 5, 034006.

Larkin, P. A. 1977. Epitaph for the concept of Maximum Sustained Yield. *Transactions of the American Fisheries Society*, 106(1), 1–11.

Lawler, A. 2010. Collapse? What collapse? Societal change revisited. *Science*, 330, 907–9.

Lee, K. 1993. *Compass and Gyroscope: Integrating Science and Politics for the Environment*. Washington, DC: Island Press.

Leiserowitz, A. A., Kates, R. W. and Parris, T. M. 2006. Sustainability values, attitudes, and behaviors: A review of multinational and global trends. *Annual Review of Environment and Resources*, 31, 413–44.

Levitus, S., Antonov, J. I., Boyer, T. P., Baranova, O. K., Garcia, H. E., Locarnini, R. A., Mishonov, A. V., Reagan, J. R., Seidov, D., Yarosh, E. S. and Zweng, M. M. 2012. World ocean heat content and thermosteric sea level change (0–2000 m), 1955–2010. *Geophysical Research Letters*, 39, L10603.

Lomborg, B. 2001. *The Skeptical Environmentalist: Measuring the Real State of the World*. New York, NY: Cambridge University Press.

Lomborg, B. and Rubin, O. 2002. The dustbin of history: Limits to growth. *Foreign Policy*, 42–44.

Lovejoy, T. 2012. A tsunami of extinction. *Scientific American*, 308, 33–34.

Luers, A. 2005. The surface of vulnerability: An analytical framework for examining environmental change. *Global Environmental Change*, 15, 214–23.

Maathai, W. 1997. "The river has been crossed": Wangari Maathai and the Mothers of the Green Belt Movement: An interview with Wangari Maathai. In: Jetter, A., Orleck, A. and Taylor, D. (eds.) *The Politics of Motherhood: Activist Voices from Left to Right*. Hanover, NH: University Press of New England.

Maathai, W. 2004. 2004 Nobel Peace Prize Lecture. The Nobel Committee.

Maffi, L. 2006. Bio-cultural diversity for endogenous development: Lessons from research, policy, and on-the-ground experiences. *Endogenous Development and Bio-Cultural Diversity*. Geneva, Switzerland.

Malthus, T. R. 1998. An essay on the principle of population, as it affects the future improvement of society with remarks on the speculations of Mr. Godwin, M. Condorcet, and other writers. Electronic Scholarly Publishing Project (www.esp.org), Lawrence, Kansas, USA. Originally published in 1798.

McAnany, P. A. and Yoffee, N. 2010. *Questioning Collapse: Human Resilience, Ecological Vulnerability, and the Aftermath of Empire*. Cambridge, UK: Cambridge University Press.

McDonald, R. I., Green, P., Balk, D., Fekete, B. M., Revenga, C., Todd, M. and Montgomery, M. 2011. Urban growth, climate change, and freshwater availability. *Proceedings of the National Academy of Sciences*, 108, 6312–17.

McKee, J. K. 2003. Reawakening Malthus: Empirical support for the Smail scenario. *American Journal of Physical Anthropology*, 122(4), 371–74.

McMichael, A. J., Butler, C. D. and Folke, C. 2003. New visions for addressing sustainability. *Science*, 302, 1919–20.

McMichael, A. J., Woodruff, R. E. and Hales, S. 2006. Climate change and human health: present and future risks. *The Lancet*, 367, 859–69.

Meadows, D. H. 2012. Envisioning a sustainable world. *Solutions*, 3, 11–14.

Meadows, D. H., Meadows, D., Randers, J. and Behrens, W. 1972. *The Limits To Growth: A Report for The Club of Rome's Project on the Predicament of Mankind*. New York: Universe Books.

Meadows, D. H., Meadows, D. and Randers, J. 1992. *Beyond the Limits: Global Collapse or a Sustainable Future*. White River Junction, VT: Chelsea Green Publishing.

Meadows, D., Randers, J. and Meadows, D. 2004. *Limits to Growth: The 30-Year Update*. White River Junction, VT: Chelsea Green Publishing.

Migdal, J. 1997. The state in society: An approach to struggles for domination. In: Migdal, J., Kohli, A. and Shue, V. (eds.) *State Power and Social Forces: Domination and Transformation in the Third World*. Cambridge, UK: Cambridge University Press.

Miles, E. L. 2009. On the increasing vulnerability of the World Ocean to multiple stresses. *Annual Review of Environment and Resources*, 34, 17–41.

Mill, J. S. 1974. *On Liberty*. London: Penguin.

Millennium Ecosystem Assessment. 2003. *Ecosystems and Human Wellbeing: A Framework for Assessment*. Washington, DC: Island Press.

Millennium Ecosystem Assessment. 2005a. *Ecosystems and Human Well-being: Synthesis*. Washington, DC: Island Press.

Millennium Ecosystem Assessment. 2005b. *Ecosystems and Human Well-being: Our Human Planet: Summary for Decision-makers*. Washington, DC: Island Press.

Millennium Ecosystem Assessment. 2005c. *Living Beyond Our Means: Natural Assets and Human Well-being*. Washington, DC: Island Press.

Moldan, B., Janoušková, S. and Hák, T. 2012. How to understand and measure environmental sustainability: Indicators and targets. *Ecological Indicators*, 17, 4–13.

Moll, P. 1993. The discreet charm of the Club of Rome. *Futures*, 25, 801–5.

Mora, C., Aburto-Oropeza, O., Ayala Bocos, A., Ayotte, P. M., Banks, S., Bauman, A. G., Beger, M., Bessudo, S., Booth, D. J., Brokovich, E., Brooks, A., Chabanet, P., Cinner, J. E., Cortés, J., Cruz-Motta, J. J., Cupul Magaña, A., Demartini, E. E., Edgar, G. J., Feary, D. A., Ferse, S. C. A., Friedlander, A. M., Gaston, K. J., Gough, C., Graham, N. A. J., Green, A., Guzman, H., Hardt, M., Kulbicki, M., Letourneur, Y., López Pérez, A., Loreau, M., Loya, Y., Martinez, C., Mascareñas-Osorio, I., Morove, T., Nadon, M.-O., Nakamura, Y., Paredes, G., Polunin, N. V. C., Pratchett, M. S., Reyes Bonilla, H., Rivera, F., Sala, E., Sandin, S. A., Soler, G., Stuart-Smith, R., Tessier, E., Tittensor, D. P., Tupper, M., Usseglio, P., Vigliola, L., Wantiez, L., Williams, I., Wilson, S. K. and Zapata, F. A. 2011. Global human footprint on the linkage between biodiversity and ecosystem functioning in reef fishes. *PLoS Biology*, 9, e1000606.

Morris, D. W. 2011. Adaptation and habitat selection in the eco-evolutionary process. *Proceedings of the Royal Society B*, 278, 2401–11.

Myers, N. 1993. Biodiversity and the precautionary principle. *Ambio*, 213, 74–79.

Myers, S. S. and Patz, J. A. 2009. Emerging threats to human health from global environmental change. *Annual Review of Environment and Resources*, 34, 223–52.

National Oceanic and Atmospheric Administration. 2014. Trends in atmospheric carbon dioxide. Earth System Research Laboratory, Global Monitoring Division, Mauna Loa, Hawaii.

National Research Council, Policy Division, Board On Sustainable Development, 1999. *Our Common Journey: A Transition Toward Sustainability*. Washington, DC: National Academies Press.

Navarrete, C. D., McDonald, M. M., Asher, B. D., Kerr, N. L., Yokota, K., Olsson, A. and Sidanius, J. 2012. Fear is readily associated with an out-group face in a minimal group context. *Evolution & Human Behavior*, 33(5), 590–93.

Nee, S. 2005. The great chain of being. *Nature*, 435, 429.

Norgaard, R. B. and Baer, P. 2005. Collectively seeing complex systems: The nature of the problem. *Bioscience*, 55, 953–60.

Norton, B. G. 2002. *Searching for Sustainability: Interdisciplinary Essays in the Philosophy of Conservation Biology*. Cambridge, UK: Cambridge University Press.

Norton, B. 2005. *Sustainability: A Philosophy of Adaptive Ecosystem Management*. Chicago, IL: University of Chicago Press.

Noss, R. F., Dobson, A. P., Baldwin, R., Beier, P., Davis, C. R., Dellasala, D. A., Francis, J., Locke, H., Nowak, K. and Lopez, R. 2012. Bolder thinking for conservation. *Conservation Biology*, 26, 1–4.

O'Hara, S. U. 1998. Economics, ethics and sustainability: redefining connections. *International Journal of Social Economics*, 25, 43–62.

O'Riordan, T. and Cameron, J. 1994. *Interpreting the Precautionary Principle*. London: Earthscan/James & James.

Oduol, W. and Kabira, W. M. 1995. The mother of warriors and her daughters: The women's movement in Kenya. In: Basu, A. (ed.) *The Challenge of Local Feminisms: Women's Movements in Global Perspective*. Boulder, CO: Westview Press.

Okereke, C. 2007. *Global Justice and Neoliberal Environmental Governance: Ethics, Sustainable Development and International Co-operation*. New York: Routledge.

Ophuls, W. 1974. The scarcity society. *The Harpers Monthly*, April, 47–52.

Ophuls, W. 2011. *Plato's Revenge: Politics in the Age of Ecology*. Cambridge, MA: MIT Press.

Orr, D. 2002. Four challenges for sustainability. *Conservation Biology*, 16, 1456–60.

Orr, D. 2012. Can we avoid the perfect storm? *Solutions*, 3. Available: www.thesolutionsjournal.com/node/1124. Accessed 5/28/2014.

Ostrom, E., Burger, J., Field, C. B., Norgaard, R. B. and Policansky, D. 1999. Revisiting the commons: Local lessons, global changes. *Science*, 284, 278–83.

Overpeck, J. T. and Cole, J. E. 2006. Abrupt change in Earth's climate system. *Annual Review of Environment and Resources*, 31(1), 1–31.

Paarlberg, R. L. 2010. *Food Politics: What Everyone Needs to Know*. Oxford, UK: Oxford University Press.

Paehlke, R. 2004. *Democracy's Dilemma: Environment, Social Equity, and the Global Economy*. Cambridge, MA: MIT Press.

Park, J., Conca, K. and Finger, M. (eds.). 2008. *The Crisis of Global Environmental Governance: Towards a New Political Economy of Sustainability*. London and New York: Routledge.

Parris, T. M. and Kates, R. W. 2003. Characterizing and measuring sustainable development. *Annual Review of Environment and Resources*, 28, 559–86.

Pereira, H. M., Navarro, L. M. and Martins, I. S. 2012. Global biodiversity change: The bad, the good, and the unknown. *Annual Review of Environment and Resources*, 37, 25–50.

Peterson, D. 2008. *Jane Goodall: The Woman Who Redefined Man*. New York: Houghton Mifflin Harcourt.

Petersen, W. 1971. The Malthus–Godwin debate, then and now. *Demography*, 8, 13–26.

Pimentel, D. 2011. Food for thought: A review of the role of energy in current and evolving agriculture. *Critical Reviews in Plant Sciences*, 30, 35–44.

Pimentel, D., Hepperly, P., Seidel, R., Hanson, J. and Douds, D. 2005. Environmental, energetic, and economic comparisons of organic and conventional farming systems. *Bioscience*, 55(7), 573–82.

Pinchot, G. 1910. *The Fight for Conservation*. New York: Doubleday, Page & Company.

Polak, P. 2008. *Out of Poverty*. San Francisco, CA: Berret-Koehler.

Ponting, C. 2007. *A New Green History of the World: The Environment and the Collapse of Great Civilizations*. London: Penguin.

Pope, D. P., Mishra, V., Thompson, L., Siddiqui, A. R., Rehfuess, E. A., Weber, M. and Bruce, N. G. 2010. Risk of low birth weight and stillbirth associated with indoor air pollution from solid fuel use in developing countries. *Epidemiologic Reviews*, 32(1), 70–81.

Princen, T. 2003. Principles for sustainability: From cooperation and efficiency to sufficiency. *Global Environmental Politics*, 3, 33–50.

Rawls, J. 1971. *A Theory of Justice*. Cambridge, MA: Belknap Press of Harvard University Press.

Redford, K. H. and Brosius, J. P. 2006. Diversity and homogenization in the endgame. *Global Environmental Change: Human and Policy Dimensions*, 16, 317–19.

Rees, W. E. 1992. Ecological footprints and appropriated carrying capacity: What urban economics leaves out. *Environment and Urbanization*, 4(2), 121–30.

Rees, W. E. and Wackernagel, M. 2013. The shoe fits, but the footprint is larger than Earth. *PLoS Biology*, 11, e1001701.

Ridgeway, S. and Jacques, P. 2013. *The Power of the Talking Stick: Indigenous Politics and the World Ecological Crisis*. Boulder, CO: Paradigm Publishers.

Robert, K.-H., Daly, H., Hawken, P. and Holmberg, J. 1997. A compass for sustainable development. *International Journal of Sustainable Development & World Ecology*, 4, 79.

Robertson, G. P. and Vitousek, P. M. 2009. Nitrogen in agriculture: Balancing the cost of an essential resource. *Annual Review of Environment and Resources*, 34, 97–125.

Robinson, W. I. 2012. Global capitalism theory and the emergence of transnational elites. *Critical Sociology*, 38, 349–63.

Rockström, J., Steffen, W., Noone, K., Persson, A., Chapin, F. S., Lambin, E. F., Lenton, T. M., Scheffer, M., Folke, C., Schellnhuber, H. J., Nykvist, B., De Wit, C. A., Hughes, T., van der Leeuw, S., Rodhe, H., Sörlin, S., Snyder, P. K., Costanza, R., Svedin, U., Falkenmark, M., Karlberg,

L., Corell, R. W., Fabry, V. J., Hansen, J., Walker, B., Liverman, D., Richardson, K., Crutzen, P. and Foley, J. A. 2009. Planetary Boundaries: Exploring the safe operating space for humanity. *Ecology and Society*, 14. Available: www.ecologyandsociety.org/vol14/iss2/art32. Accessed 5/28/2014.

Rogers, R. 1995. *The Oceans are Emptying: Fish Wars and Sustainability*. New York: Black Rose Books.

Rorty, R. 1997. Justice as a larger loyalty. *Ethical Perspectives*, 4, 139–51.

Rotmans, J. and de Vries, B. (eds.). 1997. *Perspectives on Global Change: The TARGETS Approach*. Cambridge: Cambridge University Press.

Rousseau, J. J. 1992. *Discourse on the Origin of Inequality*. Indianapolis, IN: Hacket Publishing.

Running, S. W. 2013. Approaching the limits. *Science*, 339, 1276–77.

Saith, A. 2011. Inequality, imbalance, instability: Reflections on a structural crisis. *Development and Change*, 42, 70–86.

Sandler, R. L. 2013. Environmental virtue ethics. *The International Encyclopedia of Ethics*. Hoboken, NJ: Wiley Online Library.

Schäfer, M. S. 2012. Online communication on climate change and climate politics: A literature review. *Wiley Interdisciplinary Reviews: Climate Change*, 3(6), 527–43.

Scheffer, M., Carpenter, S., Foley, J., Folke, C. and Walker, B. 2001. Catastrophic shifts in ecosystems. *Nature*, 413(October, 11), 591–96.

Sen, A. 2013. The ends and means of sustainability. *Journal of Human Development and Capabilities*, 14, 6–20.

Şengör, A. M. C., Atayman, S. and Özeren, S. 2008. A scale of greatness and causal classification of mass extinctions: Implications for mechanisms. *Proceedings of the National Academy of Sciences*, 105, 13736–40.

Simon, J. 1981. *The Ultimate Resource*. Princeton, NJ: Princeton University Press.

Simon, J. 1999. *Hoodwinking the Nation*. New Brunswick, NJ: Transaction Publishers/Cato Institute.

Simon, J. and Kahn, H. (eds.). 1984. *The Resourceful Earth: A Response to Global 2000*. New York, NY: Blackwell Press.

Smail, J. K. 2002. Remembering Malthus: A preliminary argument for a significant reduction in global human numbers. *American Journal of Physical Anthropology*, 118(3), 292–97.

Smil, V. 2000. Energy in the twentieth century: Resources, conversions, costs, uses, and consequences. *Annual Review of Energy and the Environment*, 25, 21–51.

Smil, V. 2012. *Harvesting the Biosphere: What We Have Taken from Nature*. Cambridge, MA: MIT Press.

Smith, W. J. 2012. *A Rat is a Pig is a Dog is a Boy: The Human Cost of the Animal Rights Movement*. New York: Encounter Books.

Smith, Z. A. and Farley, H. 2014. *Sustainability: If It's Everything, Is It Nothing?* London: Routledge.

Soron, D. and Laxer, G. 2006. Thematic introduction: Decommodification, democracy, and the battle for the commons. In: Laxer, G. and Soron, D. (eds.) *Not for Sale: Decommodifying Public Life* (pp. 15–37). Peterborough, Ontario: Broadview Press.

Steffen, W., Rockström, J. and Costanza, R. 2011. How defining Planetary Boundaries can transform our approach to growth. *Solutions*, 2(3). Available: www.thesolutionsjournal.com/node/935. Accessed 5/28/2014.

Stewart-Harawira, M. 2012. Returning the sacred: Indigenous ontologies in perilous times. In: Williams, L., Roberts, R. and McIntosh, A. (eds.) *Radical Human Ecology: Intercultural and Indigenous Approaches*. Farnham, Surrey, UK: Ashgate.

Stoett, P. 2012. *Global Ecopolitics: Crisis, Governance, and Justice*. Toronto: University of Toronto Press.

Stone, D. A., Allen, M. R., Stott, P. A., Pall, P., Min, S.-K., Nozawa, T. and Yukimoto, S. 2009. The detection and attribution of human influence on climate. *Annual Review of Environment and Resources*, 34, 1–16.

Storey, R. 1992. The children of Copan: Issues in paleopathology and paleodemography. *Ancient Mesoamerica*, 3(1), 161–67.

Survival International. n.d. Belo Monte Dam Background Briefing. www.survi valinternational.org/about/belo-monte-dam. Accessed 5/28/2014.

Swyngedouw, E. and Heynen, N. C. 2003. Urban political ecology, justice and the politics of scale. *Antipode*, 35, 898–918.

Tainter, J. A. 1988. *The Collapse of Complex Societies*. Cambridge, UK: Cambridge University Press.

The Club of Rome. 2006. *Founding the Club of Rome* [Online]. The Club of Rome. Available: www.clubofrome.at/peccei/clubofrome.html. Accessed 2/26/2011.

The Green Belt Movement. 2013. *Our History* [Online]. Nairobi, Kenya. Available: www.greenbeltmovement.org/who-we-are/our-history. Accessed 1/1/2013.

Thiele, L. P. 2011. *Indra's Net and the Midas Touch: Living Sustainably in a Connected World*. Cambridge, MA: MIT Press.

Tinker, G. 1996. An American Indian theological response to eco-justice. In: Weaver, J. (ed.) *Defending Mother Earth: Native American Perspectives on Environmental Justice*. Maryknoll, NY: Orbis Books.

Turner, G. M. 2008. A Comparison of the Limits to Growth with 30 Years of Reality. *Global Environmental Change*, 18, 397–411.

Turner, B. and Sabloff, J. A. 2012. Classic Period collapse of the Central Maya Lowlands: Insights about human–environment relationships for sustainability. *Proceedings of the National Academy of Sciences*, 109, 13908–14.

Turner, R. K. 2005. Sustainability: Principles and practice. In: Redclift, M. (ed.) *Sustainability: Critical Concepts in the Social Sciences, Volume 2.* Oxford, UK: Taylor & Francis.

United Nations Population Division. 2012. *World Population Prospects.* New York: United Nations.

United Nations World Commission on Environment and Development. 1987. *Report of the World Commission on Environment and Development: Our Common Future.* United Nations World Commission on Environment and Development.

Vanderheiden, S. 2008. *Atmospheric Justice: A Political Theory of Climate Change.* Oxford and New York: Oxford University Press.

Velders, G. J., Andersen, S. O., Daniel, J. S., Fahey, D. W. and McFarland, M. 2007. The importance of the Montreal Protocol in protecting climate. *Proceedings of the National Academy of Sciences*, 104, 4814–19.

Vernon, J. 2007. *Hunger: A Modern History.* Cambridge, MA: Harvard University Press.

Vitousek, P. M., Ehrlich, P. R., Ehrlich, A. H. and Matson, P. A. 1986. Human appropriation of the products of photosynthesis. *Bioscience*, 36, 368–73.

Vitousek, P. M., Mooney, H. A., Lubchenco, J. and Melillo, J. 1997. Human domination of Earth's ecosystems. *Science*, 277, 494–99.

Vollan, B. and Ostrom, E. 2010. Cooperation and the commons. *Science*, 330, 923–24.

Wackernagel, M., Schulz, N. B., Deumling, D., Linares, A. C., Jenkins, M., Kapos, V., Monfreda, C., Lo, J., Myers, N., Norgaard, R. and Randers, J. 2002. Tracking the ecological overshoot of the human economy. *Proceedings of the National Academy of Sciences (PNAS)*, 99, 9266–71.

Walker, B., Carpenter, S., Anderies, J., Abel, N., Cumming, G. S., Janssen, M., Lebel, L., Norberg, J., Peterson, G. D. and Pritchard, R. 2002. Resilience management in social-ecological systems: A working hypothesis for a participatory approach. *Conservation Ecology*, 6(1). Available: www.consecol.org/vol16/iss1/art14. Accessed 5/28/2014.

Walker, B., Gunderson, L., Kinzig, A., Folke, C., Carpenter, S. and Schultz, L. 2006. A handful of heuristics and some propositions for understanding resilience in social-ecological systems. *Ecology and Society*, 11(1). Available: www.ecologyandsociety.org/vol11/iss1/art13. Accessed 5/28/2014.

Wallerstein, I. 1974–80. *The Modern World System.* New York: Academic Press.

Waters, F. 1963. *The Book of Hopi: The First Revelation of the Hopi's Historical and Religious World-view of Life.* New York: Penguin Books.

Weiss, H. and Bradley, R. S. 2001. What Drives Societal Collapse? *Science*, 291, 609–10.

Wells, S. 2010. *Pandora's Seed: The Unforeseen Cost of Civilization.* New York: Random House.

Western, D. 2001. Human-modified ecosystems and future evolution. *Proceedings of the National Academy of Sciences (PNAS)*, 98, 5458–65.

Wiedmann, T. and Barrett, J. 2010. A review of the Ecological Footprint Indicator—perceptions and methods. *Sustainability*, 2, 1645–93.

Wilcox, M. 2010. Marketing conquest and the vanishing Indian: An Indigenous response to Jared Diamond's Guns, Germs, and Steel and Collapse. *Journal of Social Archaeology*, 10, 92–117.

Wildman, D., Uddin, M., Liu, G., Grossman, L. and Goodman, M. 2003. Implications of natural selection in shaping 99.4% nonsynonymous DNA identity between humans and chimpanzees: Enlarging genus Homo. *Proceeding of the National Academy of Sciences*, 100, 7181–8.

Williams, C. C. and Millington, A. C. 2004. The diverse and contested meanings of sustainable development. *Geographical Journal*, 170, 99–104.

Wilson, J. 1973. *Introducton to Social Movements*. New York: Basic Books.

World Commission on Environment and Development. 1987. *Our Common Future*. Oxford: Oxford University Press.

Worster, D. 1985. *Rivers of Empire: Water, Aridity, and the Growth of the American West*. New York and Newbury Park, CA: Sage, Pantheon Books.

Wu, S.-H., Ho, C.-T., Nah, S.-L. and Chau, C.-F. 2012. Global hunger: A challenge to agricultural, food, and nutritional sciences. *Critical Reviews in Food Science and Nutrition*, 54, 151–62.

Young, O. R. 2011. Effectiveness of international environmental regimes: Existing knowledge, cutting-edge themes, and research strategies. *Proceedings of the National Academy of Sciences*, 108, 19853–60.

Young, O. R. 2013. Sugaring off: Enduring insights from long-term research on environmental governance. *International Environmental Agreements: Politics, Law and Economics*, 13, 87–105.

INDEX